AF356129

Design, Technologies and Applications of High Power Vacuum Electronic Devices from Microwave to THz Band

Design, Technologies and Applications of High Power Vacuum Electronic Devices from Microwave to THz Band

Editor

Mikhail Glyavin

MDPI • Basel • Beijing • Wuhan • Barcelona • Belgrade • Manchester • Tokyo • Cluj • Tianjin

Editor
Mikhail Glyavin
Plasma Physics and High Power
Electronics
IAP RAS
Nizhny Novgorod
Russia

Editorial Office
MDPI
St. Alban-Anlage 66
4052 Basel, Switzerland

This is a reprint of articles from the Special Issue published online in the open access journal *Electronics* (ISSN 2079-9292) (available at: www.mdpi.com/journal/electronics/special_issues/DTAHPVEDMTB_electronics).

For citation purposes, cite each article independently as indicated on the article page online and as indicated below:

LastName, A.A.; LastName, B.B.; LastName, C.C. Article Title. *Journal Name* **Year**, *Volume Number*, Page Range.

ISBN 978-3-0365-2242-5 (Hbk)
ISBN 978-3-0365-2241-8 (PDF)

Contents

About the Editor

Mikhail Glyavin

Mikhail Glyavin was born in Nizhny Novgorod (then known as Gorky), Russia. After graduating from the Gorky Polytechnical Institute he has been working at the Institute of Applied Physics of the Russian Academy of Sciences, where he is engaged in the development of high-power gyrotrons. He received a Ph.D. degree and Dr.Sci. in physics from IAP RAS in 1999 and 2009, respectively. Currently he is a Deputy Director of Science and a Head of the Department of Vacuum tubes of IAP RAS. His research interests are in the field of the theoretical and experimental investigations of various gyro-devices, including gyrotrons and their application to materials processing and diagnostics of various media. Under his leadership and with his direct participation, record-breaking frequencies, powers and spectral characteristics of the radiation of gyro devices were achieved both in pulsed and continuous modes of generation; pioneering work on initiating a localized gas discharge and experiments on gas spectroscopy with record sensitivity were performed.

Preface to "Design, Technologies and Applications of High Power Vacuum Electronic Devices from Microwave to THz Band"

It is generally accepted that the 20th century was the age of electronics. Humanity has not yet decided on the topic that will dominate the current century, but many argue that it will be the century of T-rays—the century of terahertz radiation. Coherent electromagnetic radiation of subterahertz and terahertz frequency ranges with relatively high power has some specific features that make it very attractive for a wide range of fundamental and applied research areas in physics, chemistry, biology, and medicine. In particular, the enhancement of sensitivity for spectroscopic applications (electron paramagnetic resonance spectroscopy; dynamic polarization of nuclear spins in nuclear magnetic resonance spectroscopy); plasma applications (diagnostics of dense plasmas in fusion devices, the creation of compact plasma objects); biochemical applications (management of flow rate of reactions in organic chemistry, conformational changes of protein molecules); standoff detection and imaging of explosives and weapons; new medical technology; atmospheric monitoring; production of high-purity materials; deep space and special satellite communication, etc., can be mentioned.

For a long time, the intermediate position of THz waves between the microwave and optical portions of the electromagnetic spectrum did not achieve notable results in mastering powerful radiation sources and seemed to be too short in wavelength for the methods of classical vacuum electronics (due to the necessity of small-scale elements for slow-wave structures) and too low in frequency for the methods of quantum electronics (due to quant energy) to be applied.

The most widespread and frequently used devices in the "low THz" range (up to a frequency of 1.5 THz) are low-voltage and small-size backward wave oscillators (BWOs) that provide an output power of several milliwatts in the CW regime at the highest permissible frequencies. There also exist other low-voltage vacuum sources based on stimulated Cherenkov and Smith–Purcell radiation of rectilinear electron beams, as well as solid-state devices delivering sub-THz and THz radiation at a power level from hundreds of microwatts to one milliwatt; quantum cascade lasers already provide a power of hundreds of milliwatts at frequencies down to 5 THz. At the same time, the power of coherent radiation, which can be delivered by vacuum devices based on the stimulated Bremsstrahlung radiation of curvilinear electron beams (free electron lasers (FELs) and gyrotrons), can be higher by many orders of magnitude, which opens new opportunities for many applications. High-power radiation is produced in FELs and gyrotrons due to the advantage of using electrons' interaction with fast electromagnetic waves instead of slow waves in "conventional" electron devices. As a result, FELs can provide coherent and smoothly frequency-tuned radiation in the entire THz frequency region and the bands of much higher frequencies, presumably up to X-rays. However, FELs utilize ultrarelativistic electron beams and typically require huge particle accelerators for their realization. That is why FELs can be used only in specialized research centers. Unlike FELs, gyrodevices can operate with electron beams having significantly lower energies of 10–100 keV, meaning gyro-tubes are much more compact than FELs and available for many laboratories.

Currently, the above-mentioned vacuum electronic sources cover over 12 orders of magnitude in power (mW-to-GW) and 2 orders of magnitude in frequency (0.1–10 THz). The present investigations aim to develop radiation sources, including the development of new schemes of electron cyclotron masers, with record-breaking frequencies and peak and average powers. Despite the many technical limits, such as the requirement of strong operating magnetic fields for gyrodevices, mode competition, and high ohmic losses, the number of pulsed and CW radiation sources and the range of applications are increasing rapidly.

Mikhail Glyavin
Editor

 *electronics*

Article

Dependence of Irradiated High-Power Electromagnetic Waves on the Failure Threshold Time of Semiconductors Using a Closed Waveguide

Sun-Hong Min [1], Jung-Il Kim [2], Matlabjon Sattorov [3,4,5], Seontae Kim [3,4,6], Dongpyo Hong [3,4], Seonmyeong Kim [3,4], Bong-Hwan Hong [1], Chawon Park [1], Sukhwal Ma [1], Minho Kim [1], Kyo-Chul Lee [1], Yong-Jin Lee [1], Han-Byul Kwon [7], Young-Joon Yoo [4], Sang-Yoon Park [4] and Gun-Sik Park [3,4,5,*]

[1] Korea Institute of Radiological and Medical Sciences (KIRAMS), Seoul 01812, Korea; msh103@kirams.re.kr (S.-H.M.); burnn@kirams.re.kr (B.-H.H.); parknkim@kirams.re.kr (C.P.); shma@kirams.re.kr (S.M.); cleverkmh@kirams.re.kr (M.K.); kyochul@kirams.re.kr (K.-C.L.); yjlee@kirams.re.kr (Y.-J.L.)
[2] Electro-Medical Device Research Center, Korea Electrotechnology Research Institute, Ansan 15588, Korea; sky@keri.re.kr
[3] Center for THz-Driven Biomedical Systems, Department of Physics and Astronomy, College of Natural Sciences, Seoul National University, Seoul 08826, Korea; sattorovmatlabjon@gmail.com (M.S.); kimseonta@gmail.com (S.K.); dove9245594@snu.ac.kr (D.H.); smkim14@snu.ac.kr (S.K.)
[4] Center for Applied Electromagnetic Research, Advanced Institute of Convergence Technology, Suwon 16229, Korea; youngjoonyoo@snu.ac.kr (Y.-J.Y.); yoonpark77@snu.ac.kr (S.-Y.P.)
[5] R&D Department, Seoul-Teracom, Inc., Suwon 16229, Korea
[6] Young IN ACE, 1F, 60, Anyangcheondong-ro, Dongan-gu, Anyang-si 14042, Korea
[7] Home Economics Education/Biosystems and Biomedical Sciences, Korea University, Seoul 02841, Korea; dmfwl7@korea.ac.kr
* Correspondence: gunsik@snu.ac.kr

 check for updates

Citation: Min, S.-H.; Kim, J.-I.; Sattorov, M.; Kim, S.; Hong, D.; Kim, S.; Hong, B.-H.; Park, C.; Ma, S.; Kim, M.; et al. Dependence of Irradiated High-Power Electromagnetic Waves on the Failure Threshold Time of Semiconductors Using a Closed Waveguide. *Electronics* **2021**, *10*, 1884. https://doi.org/10.3390/electronics 10161884

Academic Editor: Mikhail Glyavin

Received: 18 June 2021
Accepted: 3 August 2021
Published: 6 August 2021

Publisher's Note: MDPI stays neutral with regard to jurisdictional claims in published maps and institutional affiliations.

Abstract: The failure threshold time of semiconductors caused by the impact of irradiated high-power electromagnetic waves (HPEM) is experimentally studied. A SN7442 integrated circuit (IC) is placed in an emulator with a WR430 closed waveguide and is irradiated by HPEM generated from a magnetron oscillator. The state of the SN7442 component is observed by a light-emitting diode (LED) detector and the voltage measured in the SN7442 component. As the magnitude of the electric field in the HPEM is varied from 24 kV/m to 36 kV/m, the failure threshold time falls from 195 s to 17 s with dependence of the irradiated electric field (E) on the failure threshold time (T) from $T \sim E^{-12}$ to a $T \sim E^{-6}$.

Keywords: high-power electromagnetic waves (HPEM); semiconductor; failure threshold time; microwave hardness; electromagnetic pulse (EMP) shielding

1. Introduction

High-power electromagnetic waves (HPEM) reach a target, and a sequential process of penetration and propagation occurs from the target's outer surface into its interior as the waves arrive at the electronics of the target. Modern electronic systems are principally composed of semiconductor devices, which are made primarily of silicon and gallium arsenide. These are of vital importance for the function of security systems, traffic systems and modern communication systems, and a malfunction induced by the irradiation effects of HPEM in one of these areas can cause casualties and economic disasters [1–4]. Therefore, the susceptibility of electronics to electromagnetic waves is of great interest. Theoretical predictions and experimental comparisons using a semiconductor diode have been done to inspect the dependence of the voltage pulse width on the destruction power density [5]. A failure mechanism due to a reduction on the resistance resulting from the formation of a thin conductive layer by a DC pulse is observed in a metal semiconductor field-effect transistor

(MESFET) [6]. Using an open waveguide, the destructive effects of a semiconductor, in this case, transistor-transistor-logic (TTL) and a complementary metal-oxide semiconductor (CMOS) technology-based device, caused by the impact of an electromagnetic pulse (EMP) and ultra-wideband (UWB) frequency were investigated to measure the susceptibility of electronic devices to a transient electromagnetic field threat [7–9].

With the advent of the fourth industrial era, the frequency of use of digital wireless information devices is increasing, and the demand for a new system that can quickly process large amounts of information anywhere is increasing. For this reason, the demand for the use of high frequencies higher than the frequency bandwidth currently in use is gradually increasing. The increase in the use of information devices and the dependence on information devices for the information age are also increasing. It has been reported that information equipment may malfunction or be damaged due to the effects of transient electromagnetic waves from a high-power electromagnetic wave generator operating in the frequency range of the same band. High-power microwaves (HPM) are currently being researched and developed to generate gigawatt (GW) level electromagnetic waves in a short pulse (~100 ns), with the dominant operating frequency being 1–10 GHz, but other derived frequencies exceed 30 GHz. HPM devices developed in various countries such as the United States, Russia, China, and Japan as well as several countries in Europe can be viewed as being mainly studied for military purposes. However, as described above, intentionally generated electromagnetic waves can cause serious damage to information equipment [10–18]. HPEM emulators used to measure the level at which private information devices and systems are upset by high-power transient electromagnetic waves through simulations are more effective than a gigawatt-level electromagnetic wave generator. For these basic studies, it is crucial to provide basic data for the development of an HPEM emulator.

In this paper, the dependence of irradiated HPEM on the failure threshold time is experimentally studied using a SN7442 IC in a WR430 closed waveguide emulator when the magnitude of the irradiated electric field generated from a magnetron oscillator is varied from 24 kV/m to 36 kV/m (equivalent to the range from 1.9 kW/cm^2 to 4.27 kW/cm^2).

2. The WR430 Closed Waveguide Typed Emulator

The experimental setup used to measure the dependence of the irradiated HPEM on failure threshold time using a WR430 closed waveguide type of an emulator is shown in Figure 1. The WR430 closed waveguide emulator is located between a magnetron oscillator operated at 2.46 GHz as a HPEM source and the water load. The HPEM generated from the magnetron oscillator propagates along the WR430 waveguide and is irradiated into the semiconductor located at the bottom of the emulator [19–23].

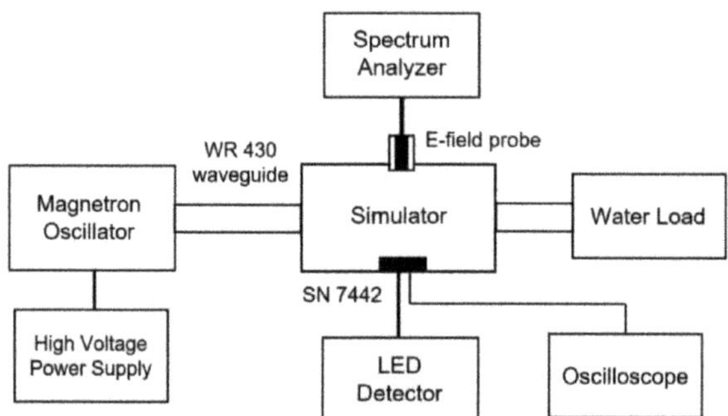

Figure 1. Experimental setup of the emulator equipped with a magnetron, and various fixtures.

The SN7442 IC is used in the experiment and functions to form a decimal code using a binary code. This is connected to a light-emitting diode (LED) detector which is turned on by the SN7442 IC, with the state of the SN7442, such as upset and destruction states, directly detected. To observe the time dependence caused by irradiated electric field, the voltage in the SN7442 IC is measured using an oscilloscope during the irradiation process. Using an E-field probe located on the upper side of the SN7442 IC, the magnitude of the irradiated electric field is measured.

The irradiated electric field is measured from the power coupling level between port 1 and port 2, as shown in Figure 2, using an E-field probe located on the upper side of the emulator. The E-field probe consisted of Teflon with a ϕ value of 4.1 mm and brass with a ϕ value of 1.26 mm. The relationship between the power coupling level and the magnitude of the irradiated electric field is expressed as follows. The charge induced in the central conductor of the E-field probe by the propagated electric field of E_0 along the waveguide is expressed as $Q_{in} = \int D \cdot \hat{n} da = \varepsilon_o E_o A_{eff}$. The area of the induced charge is modified to the effective area, A_{eff}, due to the modification of the electric field near the central conductor of the E-field probe. The area of the induced charge is expressed as $A_{eff} = K_p A_0$. Here, A_0 is the normal surface area of the central conductor in the E-field probe and K_p is the area multiplication factor, which indicates the entire modified area of the induced charge in the central conductor of the E-field probe. The induced current is $I(t) = \varepsilon_o \omega E_o K_p A_o e^{j\omega t}$ and the observed power in port 2 is expressed as $P_{ob} = \frac{1}{2}|I|^2 Z_o = \frac{Z_o}{2}\left(\varepsilon_o \omega E_o K_p A_o\right)^2$, where Z_0 is the impedance of the E-field probe and ω is the HPEM frequency. This is re-expressed as

$$P_{ob} = \frac{2\omega \mu_o Z_o P_{in}}{k_z ab}\left(\varepsilon_o \omega K_p A_o\right)^2 \tag{1}$$

where P_{in} is the input power at port 1. Using the relationship between the observed power and the input power, the area multiplication factor is calculated as follows:

$$K_p = \frac{1}{\varepsilon_o \omega A_o}\sqrt{\frac{k_z ab P_{ob}}{2\omega \mu_o Z_o P_{in}}} \tag{2}$$

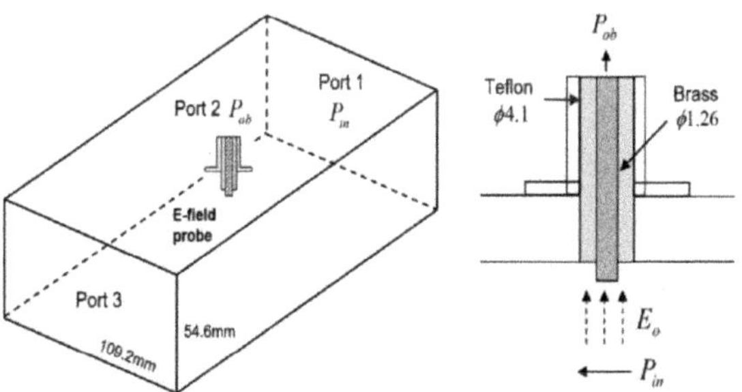

Figure 2. Design of the electric field probe.

From the measurement, P_{ob}/P_{in} is measured and found to be −38.2 dB, and K_p is calculated as 26. Using K_p and P_{ob}, the irradiated electric field is calculated where $Z_0 = 50\ \Omega$.

$$E_o = \frac{1}{\varepsilon_o \omega K_p}\sqrt{\frac{2P_{ob}}{Z_o}} \tag{3}$$

The electric field probe used to measure the electric field to be applied to the experimental semiconductor was designed using Microwave Studio, an electromagnetic (EM) simulation code, and the area multiplication factor equation discussed in relation to the relationship between the electromagnetic wave strength measured by the probe and the electric field. As shown in Figure 3, the emulator to be used in the experiment is composed of a closed waveguide. High-power electromagnetic waves generated by the magnetron, a high-power electromagnetic wave generator, are applied to Port 1 and pass through Port 3, with some of these waves detected by a probe located in Port 2. The detected output strength can be measured using a spectrum analyzer. At this time, the cross-section of the used electric field probe has a structure consisting of brass with a diameter of 1.26 mm in the center and Teflon with a diameter of 4.1 mm around it, as shown in Figure 3. Because the frequency component of the high-power electromagnetic wave generated by the magnetron is 2.45 GHz, a standard WR 430 waveguide served as the closed waveguide. The cross-sectional size of this waveguide is 54.6 mm × 109.2 mm. When an electromagnetic wave with a certain level of output strength is made to pass by using the electromagnetic simulation code, the output strength detected by the probe to be used in the experiment is measured, and the characteristics of the experimental devices can be compared.

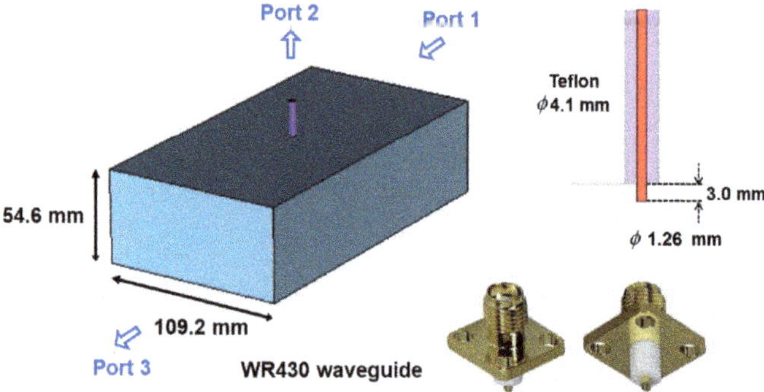

Figure 3. Simulation model for the electric-field probe design.

The electric field distribution obtained using the EM simulation code is shown in Figure 4. When the electromagnetic wave generated by the magnetron passes through the WR 430 waveguide, it shows the electric field distribution of the TE_{10} mode, the most basic mode of the waveguide. The experimental semiconductor is located at the bottom of the waveguide, and the electric field probe is located directly above the semiconductor. This configuration allows the electric field applied to the semiconductor to be measured.

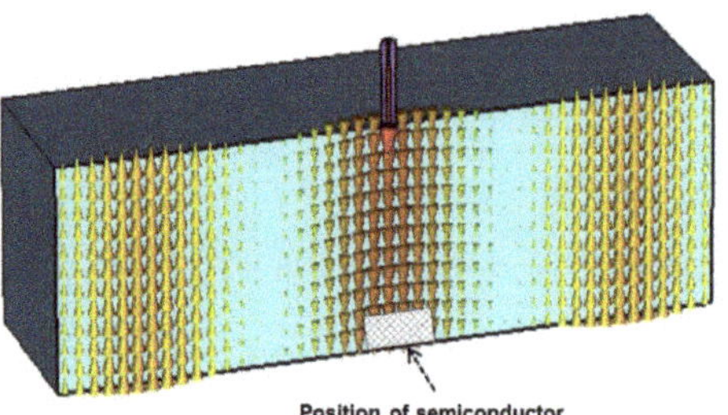

Figure 4. Result of the electric-field distribution using the EM simulation code.

In order to measure the electric field strength applied to the semiconductor accurately, the ratio of the output strength detected by the probe from the waveguide must be measured accurately. Figure 5 shows the ratio of the measured output strength using a simulation for this purpose. This is to secure the reliability and reproducibility of the measurement and to stabilize the field-to-probe binding for a fixed length.

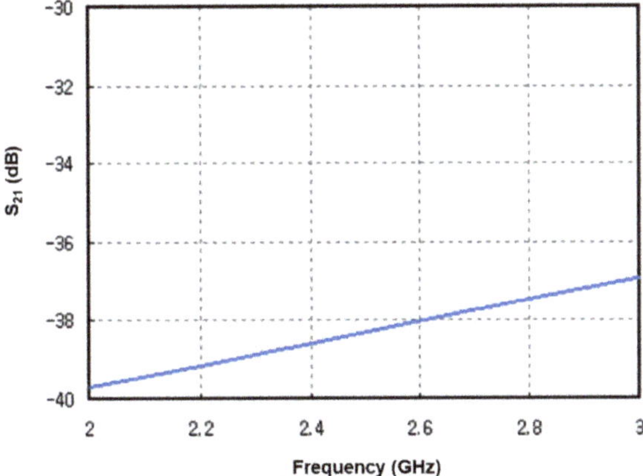

Figure 5. Ratio of the power intensity coupled from the waveguide to the electric field probe.

Figure 6 depicts the measurement line used to measure the strength of the electric field applied to the semiconductor via a simulation. Regarding the electric field in the area where the probe is located, it was determined to be slightly higher than the electric field strength at the bottom surface due to the influence of the center conductor of the probe; the influence of the probe disappears in the area outside the influence of the probe, meaning that only the electric field affected by the electromagnetic wave passing through the waveguide is measured.

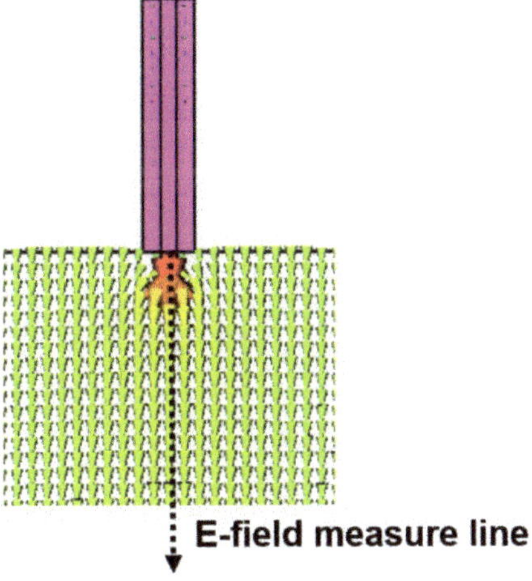

Figure 6. Ratio of the power intensity coupled from the waveguide to the electric field probe.

Figure 7 shows the result of the measurement of the electric field strength in the area where the probe and semiconductor are located via the aforementioned simulation. The position of length = 0 is the position of the upper surface of the waveguide where the probe is located, and the position of length = 50 indicates the bottom of the waveguide where the semiconductor is located. It can be seen that the strength of the electric field in the region where the probe is located is greater than that in the region where the semiconductor is located due to the central conductor of the probe. However, because the influence of the probe does not affect the lower part where the semiconductor is located, it can be confirmed that an electric field of a certain size is applied to the semiconductor. The input strength of the electromagnetic wave used to obtain this result was 1 W, and the measured electric field strength was 552 V/m. The strength of the electric field at the input strength of 1 W obtained through the simulation can be compared with the theoretical value by examining the relationship between the input strength and the electric field. Given that the result is consistent with the theoretical value, the simulation result is reliable, and through this, the electric field probe was designed.

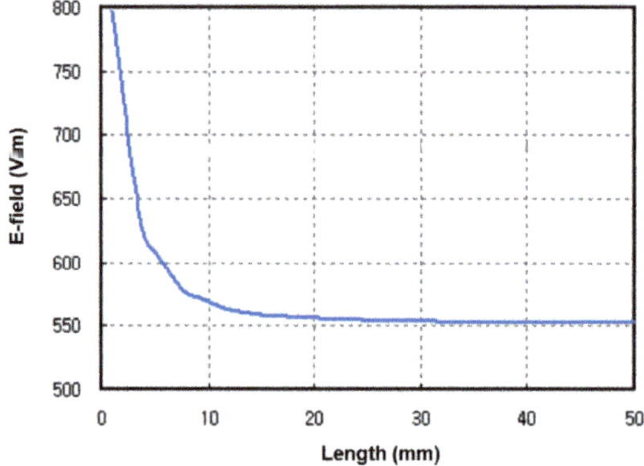

Figure 7. Electric field in the area where the probe and semiconductor are located.

3. Experimental Setup

Figure 8 shows the configuration of the experimental apparatus used to obtain and analyze the destruction data of semiconductors exposed to intentionally generated high-power electromagnetic waves (HPEM). In order to create high-power transient electromagnetic waves intentionally, the magnetron, which is the HPEM source, is located on the left side of the experimental device, and the high-power electromagnetic waves generated from the magnetron are connected to the launcher and transmitted to the WR430 waveguide. They pass through the emulator and propagate to the termination point located at the right end. In order to eliminate heat generation due to the high-power electromagnetic waves, the termination point is structured to perform refrigeration using circulating cooling water.

The magnetron, a high-power electromagnetic wave generator, generates electromagnetic waves by applying voltage to the cathode that generates an electron beam and the anode that accelerates the generated electron beam. 4.5 V and 10A are applied to the cathode and voltage of 4 kV or more is applied to the anode. In this way, the magnetron can generate high-power electromagnetic waves at the desired output strength. A high voltage probe was used to measure the voltage applied to the magnetron from a high voltage power supply. It is equipped with an oscilloscope available to measure the voltage applied to the anode. Moreover, it can find the strength value of the electric field applied to the semiconductor through the electric field probe located on the upper part of the semiconductor. By coupling electromagnetic waves of a certain intensity level, it was possible to measure the electric field applied to the semiconductor with the measured output intensity using a spectrum analyzer.

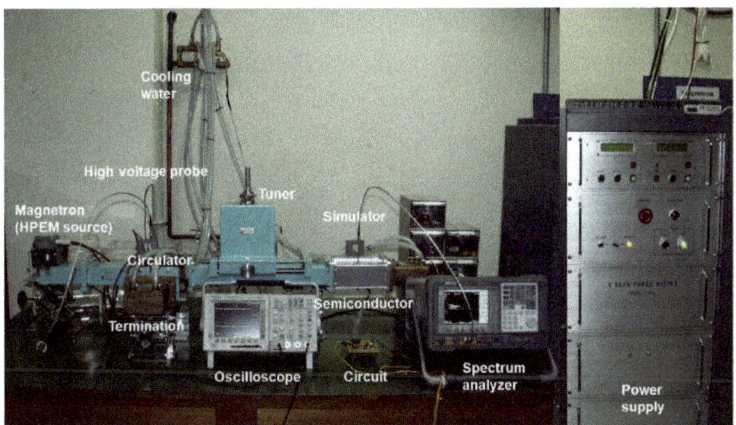

Figure 8. An emulator and a magnetron-based high-power transient electromagnetic experiment device.

Two methods were used to check for certain phenomena directly, such as upset/destruction phenomena that occur when high-power electromagnetic waves are applied to a semiconductor. First, in order to verify the characteristics of the high-power electromagnetic waves, we put the semiconductor in the experiment in the emulator and used a connection line to construct an LED circuit operated by the semiconductor while in the emulator. If an abnormality occurs in the semiconductor, the LED circuit operated by the semiconductor is abnormal. The upset/destruction characteristics can then be measured. Another measurement method used is to measure the voltage applied to the semiconductor in the emulator and to measure the applied voltage of the semiconductor, which changes when a high-power transient electromagnetic wave is applied. This allows its characteristics to be measured.

The emulator used in this experiment consists of a closed waveguide that supplies a uniform electric field to the experimental semiconductor. The lower surface of the waveguide has a groove for inserting the experimental semiconductor, and the electric field probe for measuring the electric field applied to the semiconductor is located on the upper surface at the same position so that the electric field used in the experiment can be measured.

Figure 9 shows the emulator arrangement, consisting of a closed waveguide. The emulator is located between two waveguides, which are in this case WR 430 waveguides that allow high-power electromagnetic waves generated by the magnetron to propagate through a certain area. The electric field probe is located at the top of the emulator, and an attenuator is connected to the probe to prevent strong electromagnetic waves from entering the spectrum analyzer.

In the emulator, a semiconductor for measuring the effects of electromagnetic waves is positioned, and the LED detector circuit driven due to the characteristics of the semiconductor is connected with a connection line so that the effect of the high-power electromagnetic waves on the semiconductor can be checked in real time. Moreover, as shown in the figure on the right side of Figure 9, the voltage applied to the semiconductor for the experiment was measured to determine how the applied electromagnetic wave affects the applied voltage characteristics of the semiconductor.

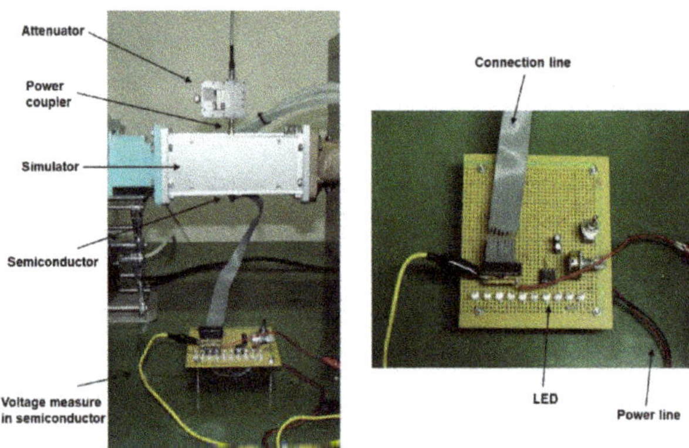

Figure 9. An emulator setup consisting of a closed waveguide.

Figure 10 shows the circuit diagram of an LED detector configured to measure the change in the characteristics of semiconductors exposed to electromagnetic waves in the experiment. Three semiconductors (SN 7442, SN7490, NE555) were used to construct this circuit. The semiconductor exposed to the double electromagnetic waves is the SN7442 type; it receives a binary input signal and converts it to a decimal number and then turns on ten LEDs one after the other. When electromagnetic waves are not applied, the ten LEDs light up at regular intervals, and when electromagnetic waves are applied, the situation in which the LEDs are turned on without any tendency due to changes in certain characteristics (such as an upset event) can be directly observed through the LEDs. To measure the voltage applied to the semiconductor, the voltage V18 (the voltage signal required to operate LED 1) was measured using an oscilloscope, as shown in the Figure 10.

Figure 10. Circuit diagram of the LED detector.

Figure 11 shows the LED detector circuit, which was devised based on the circuit diagram described in Figure 10. The characteristics of the semiconductor used in the circuit diagram are as follows. NE555 is applied to a non-stable multivibrator; this component is widely used as a pulse signal source for timer circuits and counter circuits. It can be used in TTL or CMOS circuits because it can operate in the DC voltage range of 4~15 V, and its rated current capacity is as large as 200 mA. The circuit used in the experiment supplies pulses to the counter (IC) SN7490.

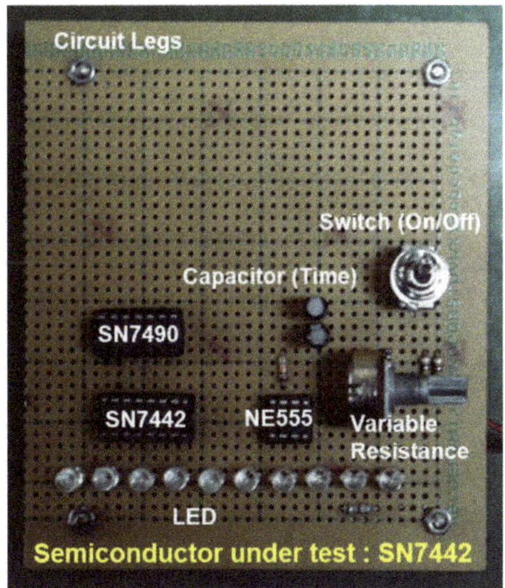

Figure 11. Circuit of the LED detector.

The SN7490 component in Figure 12 is an IC with built-in binary and counter components. The clock pulse input of the binary counter is input A (#14) and the output is Q_a (#12), and the clock pulse input of the pentagram counter is input B (#1) The outputs are Q_b (#9), Q_c (#8), and Q_d (#11). Therefore, to use this as a BCD decimal counter, input A (#14) must be used as the clock pulse input, and Q_a (#12) and input B (#1) must be connected before use. The outputs are Q_a (#12), Q_b (#9), Q_c (#8), Q_d (#11), but Q_a is the lowest digit and Q_d is the highest digit.

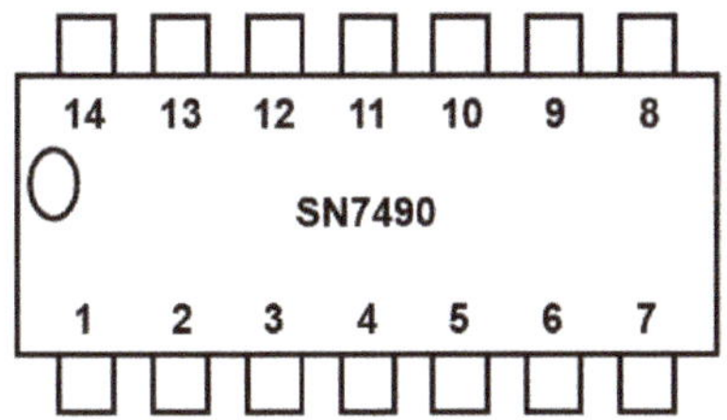

#1 pin	input A	#8 pin	Q_c
#2 pin	$R_0(1)$	#9 pin	Q_b
#3 pin	$R_0(1)$	#10 pin	GND
#4 pin	NC	#11 pin	Q_d
#5 pin	V_{cc}	#12 pin	Q_a
#6 pin	$R_g(1)$	#13 pin	NC
#7 pin	$R_g(2)$	#14 pin	input B

Figure 12. Characteristics of the SN7490 component.

The SN7442 component in Figure 13 converts (decodes) the BCD output of the decimal counter (SN7490) to decimal and operates the corresponding LED. Therefore, the output LED turns on sequentially from No. 0 to No. 9. Table 1 below explains Q_a~Q_d.

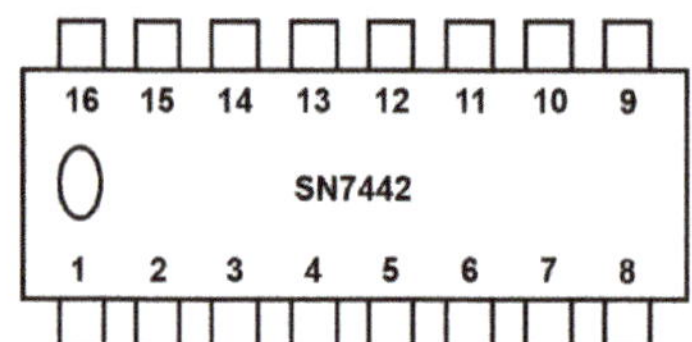

#1 pin	Decimal output 0	#9 pin	Decimal output 7
#2 pin	Decimal output 1	#10 pin	Decimal output 8
#3 pin	Decimal output 2	#11 pin	Decimal output 9
#4 pin	Decimal output 3	#12 pin	BCD input (Q_d)
#5 pin	Decimal output 4	#13 pin	BCD input (Q_c)
#6 pin	Decimal output 5	#14 pin	BCD input (Q_b)
#7 pin	Decimal output 6	#15 pin	BCD input (Q_a)
#8 pin	GND	#16 pin	Power V_{cc}

Figure 13. Characteristics of the SN7442 component.

Table 1. Input Pulses vs. Output for Q_a~Q_d.

Input Pulse	Output			
	Q_d	Q_c	Q_b	Q_a
0	0	0	0	0
1	0	0	0	1
2	0	0	1	0
3	0	0	1	1
4	0	1	0	0
5	0	1	0	1
6	0	1	1	0
7	0	1	1	1
8	1	0	0	0
9	1	0	0	1

4. Experimental Results

Figure 14a shows the LED detector in operation when a high-power electromagnetic wave is applied to a semiconductor. It can be observed that the operation of the LED changes as the intensity of the electromagnetic wave increases. Figure 14b presents the measurement of the voltage signal applied to the semiconductor with an oscilloscope when high-power electromagnetic waves are applied to the semiconductor.

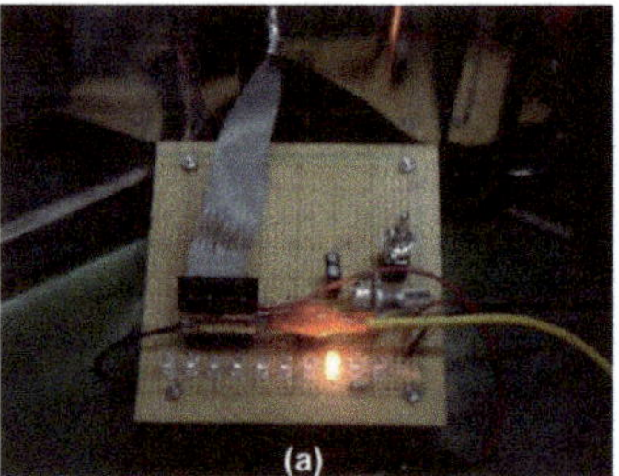

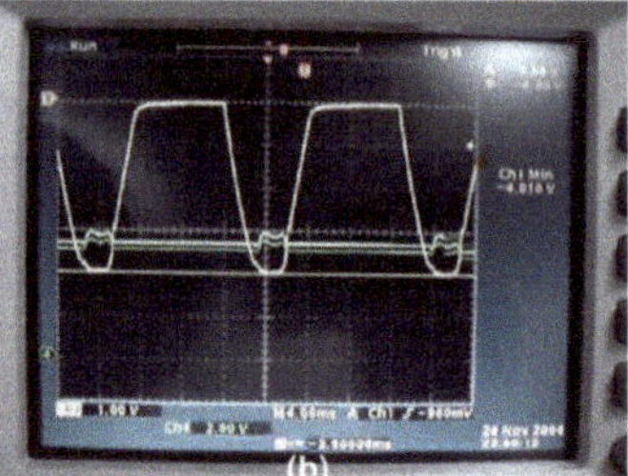

Figure 14. HPEM emulator experiment: (**a**) LED detector in operation, and (**b**) semiconductor voltage being measured.

When the high-power electromagnetic waves were applied to the semiconductor, the voltage signal applied to the semiconductor was measured, as shown in Figure 15a–d. Figure 15a shows the beam voltage and the voltage signal applied to the semiconductor before the high-power electromagnetic wave is applied, and Figure 15b is the beam voltage and the voltage signal applied to the semiconductor when the high-power electromagnetic wave is applied. Figure 15c indicates the beam voltage and the voltage signal applied to the semiconductor after a certain period of time has elapsed after the high-power electromagnetic wave is applied, as shown by the voltage signal.

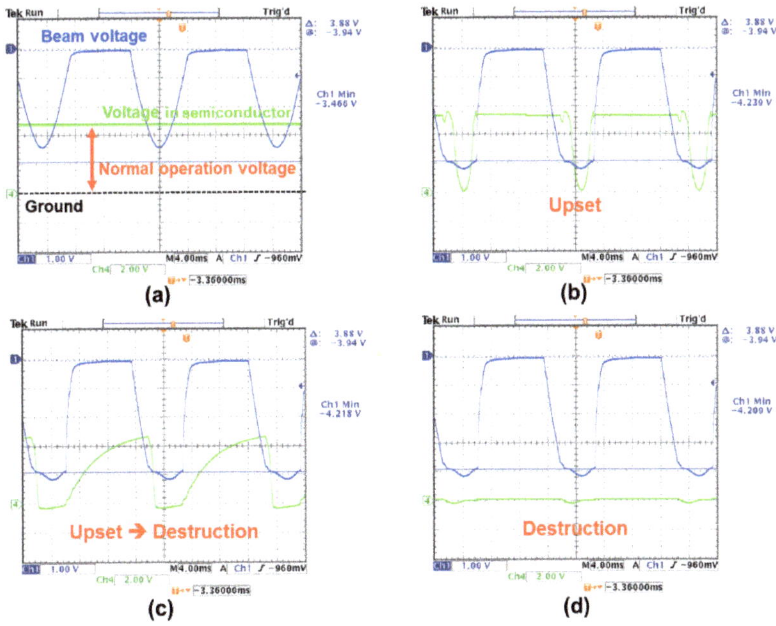

(a)

(b)

(c)

(d)

Figure 15. Voltage behavior in the semiconductor: (**a**) beam voltage before the high-power electromagnetic wave is applied and voltage signal applied to the semiconductor, (**b**) beam voltage when high-power electromagnetic waves start to be applied and voltage signals are applied to the semiconductor, (**c**) beam voltage and voltage signal applied to the semiconductor after a certain period of time after the high-power electromagnetic wave is applied, and (**d**) beam voltage and voltage signal applied to the semiconductor after the destruction of the semiconductor occurs after the application of the high-power electromagnetic waves.

As shown in Figure 15a, the measured part shown in blue represents the high voltage signal applied to the anode to drive the magnetron, and the measured part shown in green indicates the voltage applied to the semiconductor. This is a measure of the signal. As shown in Figure 15a, a certain voltage signal is applied to the semiconductor before the high-power electromagnetic wave is applied, and it operates normally. When the electromagnetic wave is applied, a situation in which the voltage signal of the semiconductor goes to 0 at the part where the electromagnetic wave is applied is noted, as shown in Figure 15b. This shows the conversion process. In Figure 15c, when high-power electromagnetic waves are applied for a certain period of time, the semiconductor is affected at the region where the electromagnetic waves are not applied such that the short-circuiting time becomes longer and the influence on the semiconductor lasts for a longer time. Figure 15d shows the time when the voltage signal is measured as zero at all times when the semiconductor is destroyed, and this indicates that the operation of the semiconductor does not recover again after the destruction.

Figure 16 shows the relationship between the destruction time of the semiconductor and the change in the output intensity generated by the magnetron, that is, the change in the electric field applied to the semiconductor. When the applied electric field changes from 24 kV/m to 36 kV/m, it can be observed that the destruction time decreases from 200 s to 15 s. Moreover, the greater the strength of the electric field applied to the semiconductor, that is, the greater the energy received by the semiconductor, the faster it is destroyed, and the relationship between the destruction time and the strength of the electric field can be divided into three regions based on a certain region. When region 1 is the region, when an electric field with low intensity is applied, the destruction time is proportional to -12 power of the electric field, and when region 2 is the region, when an electric field of about medium intensity is applied, the destruction time is proportional to the power of -10. Finally, region 3 is a region where a rather strong electric field is applied, and the destruction time is in proportion to -6 power of the electric field. This shows that the greater the electric field strength, the weaker the effect of the destruction time. Eventually, destruction of electronic devices exposed to high-power transient electromagnetic waves (HPEM) means physical damage and defines a case that can only be recovered through hardware replacement. The destruction time can be thought of in relation to the electric field, and it can be inferred that it is visually defined as a functional relationship with a tendency between the destruction time and threshold electric field, as shown in Figure 16.

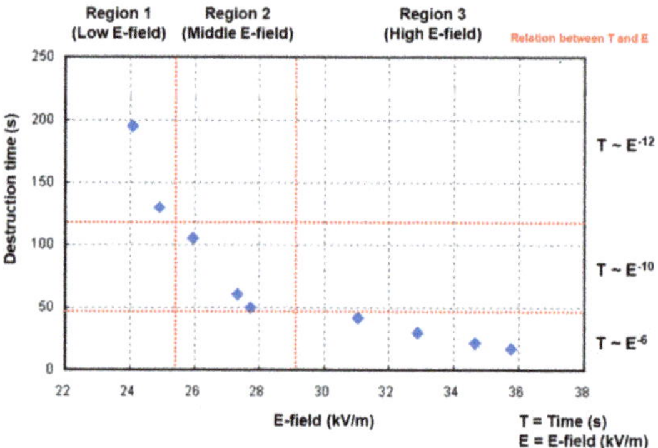

Figure 16. Destruction time as a function of the E–field.

Figure 17a,b are, correspondingly, graphs of one pulse energy according to the strength of the electric field and an electric graph according to the strength of the field. At a greater intensity of the electric field, that is, when the energy of one pulse is greater, less total energy that destroys the semiconductor can enter.

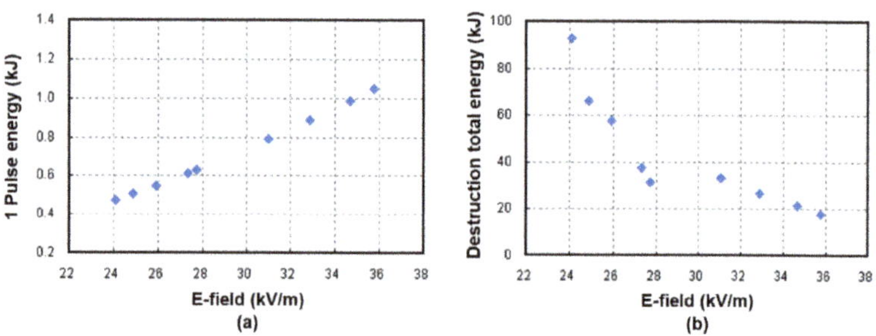

Figure 17. Destruction energy as a function of E-field; (a) one Pulse energy as a function of the E-field, (b) destruction total energy as a function of the E-field.

Figure 18 presents a comparison of the pulse energy according to the electric field strength with other experimental values, specifically those from J.H. Mcadoo and D.C Wunsch. In the Wunsch case, the energy of one pulse is shown when a DC pulse is applied using a line directly connected to the semiconductor. The experiment using the waveguide was devised by M. Camp with TTL/CMOS using an open waveguide. As shown in the M. Camp case, the electromagnetic wave used has a strong electric field of 1300 kV/m, but with a very short pulse. In this experiment with a closed waveguide, the strength of the electric field is weak, but a rather long pulse is used. In this experiment, the energy level is high.

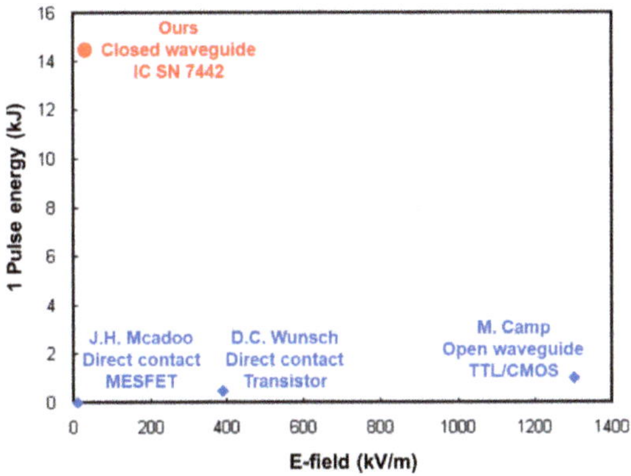

Figure 18. Pulse energy comparison according to the electric field strength.

Table 2 shows a comparison of each of the experiments related to transient electromagnetic waves. From the measured electric field strength, the condition of the electromagnetic wave required for the destruction of the semiconductor shows that the pulse width is short, whereas the strong electric field can cause destruction with little energy.

Table 2. Comparison of Experiments Related to the HPEM Effect.

Author	Semiconductor under Test	Experimental Method	Measured E-Field	Result
Wunsch, D.C. Ref. [5]	Transistor	Direct contact	Destruction 500 kW/cm^2 (390 kV/m) DC pulse~0.1 μs	Relationship between the pulse width and the threshold power
Mcadoo, J.H. Ref. [6]	MESFET	Direct contact	Destruction 400 W/cm^2 (11 kV/m) DC pulse~0.086 μs	Relationship between the DC pulse rise time and the damage voltage
Lovetri, J. Ref. [24]	PC	Horn antenna	Reset 32.9 mW/cm^2 (100 V/m)	The electric field region measurement for reset and power down scenarios
Camp, M. Ref. [7]	TTL CMOS	Open waveguide	Destruction 5.57 MW/cm^2 (1300 kV/m), 0.18 μs	Relationship between the E-field strength and the destruction rate
Ours	IC SN 7442	Closed waveguide	Destruction 31 kV/m (3.17 kW/cm^2), 4 ms	Relationship between the E-field strength and the destruction time

5. Discussion

In order to secure and analyze the threshold data of information devices exposed to high-power electromagnetic waves (HPEM), we reconfigured the HPEM environment to fit a suitable experimental environment for this. Through such a simulation, we intended to develop the emulator capable of measuring the level at which private information devices and systems are upset and/or destroyed by high-power transient electromagnetic waves, with the same effect considered accurately, as opposed to a study of an electromagnetic wave generator with the highest output at the gigawatt level. An electromagnetic wave generator with output power in the range of 1 W to 1 kW can be used. This basic research provides basic data for the development of emulators.

It is important to design an emulator consisting of a closed conduit that creates a uniform electric field for the investigations of the semiconductor component threshold characteristics. Here, it was designed/manufactured and used, and a magnetron with an operating frequency of 2.45 GHz and maximum output of 1 kW was used as the HPEM electromagnetic wave source. A breakdown/reset, destruction experiment, well known to influence the E-field, was designed with the corresponding control unit, a LED, and a power supply to measure the threshold while monitoring the operational status of the waveguide and an electronic device connecting the emulator and the electromagnetic wave in real time. In particular, this study analyzed and collected data by investigating the influence of semiconductor devices on the operating frequency, pulse duty, and electric field strength and direction.

As described above, as preliminary research on EMP protection methods and attacks, experimental research based on the emulator can be very important. The present study investigated the interaction between transient electromagnetic waves and semiconductor devices and examined conditions such as failures, damage, and malfunctions to determine the maximum degree of exposure to transient electromagnetic waves of sensitive communication devices. A database was compiled by experimenting with various types of semiconductor devices. Through a similar experiment on the communication equipment as a system, it can be possible to define the maximum transient electromagnetic wave allowed for all communication equipment. On the other hand, through the basic forms of this research, various hardening methods, such as shielding to minimize the effects of transient electromagnetic waves on semiconductors, can be developed.

6. Conclusions

In order to develop a technology to protect information equipment in response to the generation of high-power transient electromagnetic waves, a study of the effects on information equipment due to the generation of high-power transient electromagnetic waves must first be done first, which is what is presented here. In this study, a simulation was performed using a magnetron that generates an electromagnetic wave with an output of 1 kW at a frequency of 2.45 GHz. This was done in the laboratory of the authors in order to analyze the destructive effects by the E-field of a semiconductor component. Unlike previous, related studies, a closed waveguide was used for accurate experimental reproducibility and precise electric field measurements and control, and the data were collected and analyzed by measuring the threshold electric field affecting the semiconductor device. This research will provide basic data for the development of an HPEM emulator and the development of protection technology for information devices to secure them against high-power transient electromagnetic waves.

Author Contributions: Conceptualization, S.-H.M. and J.-I.K.; methodology, S.-H.M., J.-I.K. and C.P.; software, M.K. and S.M.; validation, K.-C.L. and Y.-J.L.; formal analysis, M.S., S.K. (Seontae Kim) and D.H.; investigation, S.K. (Seonmyeong Kim); resources, Y.-J.Y. and S.-Y.P.; data curation, S.-H.M., J.-I.K., M.S., H.-B.K. and C.P.; writing—original draft preparation, S.-H.M. and J.-I.K.; writing—review and editing, S.-H.M., J.-I.K., C.P., M.K., S.M., B.-H.H. and G.-S.P.; visualization, S.-H.M. and J.-I.K.; supervision, S.-H.M., J.-I.K., B.-H.H. and G.-S.P.; project administration, S.-H.M., J.-I.K., B.-H.H. and G.-S.P.; funding acquisition, S.-H.M., B.-H.H. and G.-S.P. All authors have read and agreed to the published version of the manuscript.

Funding: This work was supported by the National Research Foundation of Korea (NRF) grant funded by the Korea government (MSIP) (No. NRF—2021M2E8A1038938, No. NRF—2021R1F1A1048374, and No. NRF—2016R1A3B1908336). This study was also supported by a grant of the Korea Institute of Radiological and Medical Sciences (KIRAMS), funded by the Ministry of Science and ICT (MSIT), Republic of Korea. (No. 50051-2021, 50623-2021, and 50532-2021).

Data Availability Statement: The data that support the findings of this study are available from the corresponding author upon reasonable request.

Conflicts of Interest: The authors declare no conflict of interest.

References

1. Olesen, H.L. *Radiation Effects on Electronic Systems*, 1st ed.; Springer: New York, NY, USA, 1966.
2. Taylor, C.D.; Giri, D.V. *High-Power Microwave Systems and Effects*, 1st ed.; CRC Press: Boca Raton, FL, USA, 1994.
3. Lee, K.S.H. *EMP Interaction Principles, Techniques, and Reference Data*, 1st ed.; Springer: Berlin/Heidelberg, Germany, 1986.
4. James, B.; John, A.S.; Edl, S. *High Power Microwaves*, 3rd ed.; CRC Press: Boca Raton, FL, USA, 2016.
5. Wunsch, D.C.; Bell, R.R. Determination of threshold failure levels of semiconductor diodes and transistors due to pulse voltages. *IEEE Trans. Nucl. Sci.* **1968**, *15*, 244–259. [CrossRef]
6. McAdoo, J.H.; Bollen, W.M.; Catoe, W.; Kaul, R. Broad-band electromagnetic radiation damage in GaAs MESFETs. In Proceedings of the Microwave and Millimeter-Wave Monolithic Circuits Symposium 1992, San Diego, CA, USA, 10–13 May 1992; pp. 205–208.
7. Camp, M.; Garbe, H.; Nitsch, D. Influence of the technology on the destruction effects of semiconductors by impact of EMP and UWB pulses. In Proceedings of the 2002 IEEE International Symposium on Electromagnetic Compatibility, Minneapolis, MN, USA, 19–23 August 2002; pp. 87–92.
8. Camp, M.; Gerth, H.; Garbe, H.; Haase, H. Predicting the breakdown behavior of microcontrollers under EMP/UWB impact using a statistical analysis. *IEEE Trans. Electromagn. Compat.* **2004**, *46*, 368–379. [CrossRef]
9. Sabath, F.; Giri, D.V.; Rachidi, F.; Kaelin, A. *Ultra-Wideband, Short Pulse Electromagnetics 9*, 1st ed.; Springer: Berlin/Heidelberg, Germany, 2010.
10. Min, S.H.; Kwon, O.; Sattorov, M. Effects on electronics exposed to high-power microwaves on the basis of a relativistic backward wave oscillator operating on the X-band. *J. Electromagn. Waves Appl.* **2017**, *31*, 1875–1901. [CrossRef]
11. Kesari, V.; Basu, B.N. *High Power Microwave Tubes: Basics and Trends*, 1st ed.; Morgan & Claypool Publishers: San Rafael, CA, USA, 2018.
12. Zhen, K.; Shanwei, L.; Yan, Z.Y. Research on damage of intense electromagnetic pulse to radar receiving system. In Proceedings of the 2012 5th Global Symposium on Millimeter-Waves, Harbin, China, 27–30 May 2012; pp. 458–461.
13. Military Standard. *High-Altitude Electromagnetic Pulse (HEMP) Protection for Ground-Based C4I Facilities Performing Critical, Time-Urgent Missions, Part 1 Fixed Facilities*; MIL-STD-188-125-1; Department of Defense: Washington, DC, USA, 2005.
14. Wilson, C. High Altitude Electromagnetic Pulse (HEMP) and High Power Microwave (HPM) Devices: Threat Assessments; CRS Report for Congress, 26 March 2008. 26 March 2008. Available online: https://www.everycrsreport.com/reports/RL32544.html (accessed on 2 August 2021).
15. Giri, D.V.; Tesche, F.M. Classification of Intentional Electromagnetic Environments (IEME). *IEEE Trans. Electromagn. Compat.* **2004**, *46*, 322–328. [CrossRef]
16. Kim, I.; Kovitz, J.M.; Rahmat-Samii, Y. Enhancing the Power Capabilities of the Stepped Septum Using an Optimized Smooth Sigmoid Profile. *IEEE Trans. Antennas Propag.* **2014**, *56*, 16–42. [CrossRef]
17. Giri, D.V.; Hoad, R.; Sabath, F. *High-Power Electromagnetic Effects on Electronic Systems*, 1st ed.; Artech House: Boston, MA, USA, 2020.
18. Zhang, J.; Zhang, D.; Fan, Y. Progress in narrowband high-power microwave sources. *Phys. Plasmas.* **2020**, *27*, 010501. [CrossRef]
19. Kim, J.I.; Won, J.H.; Park, G.S. Numerical Study of a 10-Vane Strapped Magnetron Oscillator. *J. Korean Phys. Soc.* **2004**, *44*, 1229–1233.
20. Jung, S.S.; Jin, Y.S.; Kim, J.I. Three-Dimensional Particle-in-Cell Simulations of a Strapped Magnetron Oscillator. *J. Korean Phys. Soc.* **2004**, *44*, 1250–1255.
21. Kim, J.I.; Won, J.H.; Ha, H.J. Three-Dimensional Particle-in-Cell Simulation of 10-Vane Strapped Magnetron Oscillator. *IEEE Trans. Plasma Sci.* **2004**, *32*, 2099–2104. [CrossRef]
22. Kim, J.I.; Won, J.H.; Park, G.S. Electron prebunching in microwave magnetron by electric priming using anode shape modification. *Appl. Phys. Lett.* **2005**, *86*, 171501. [CrossRef]
23. Kim, J.I.; Won, J.H.; Park, G.S. Reduction of noise in strapped magnetron by electric priming using anode shape modification. *Appl. Phys. Lett.* **2006**, *88*, 221501. [CrossRef]
24. LoVetri, J.; Wilbers, A.T.M.; Zwamborn, A.P.M. Microwave interaction with a personal computer: Experiment and modelling. In Proceedings of the 13th International Zurich Symposium Technical Exhibition on Electromagnetic Compatibility, Zurich, Switzerland, 16–18 February 1999; pp. 203–206.

 *electronics*

Article

Design and Preliminary Experiment of W-Band Broadband TE$_{02}$ Mode Gyro-TWT

Xu Zeng [1,*], Chaohai Du [2,*], An Li [1], Shang Gao [1], Zheyuan Wang [1], Yichi Zhang [1], Zhangxiong Zi [1] and Jinjun Feng [1]

[1] National Key Laboratory of Science and Technology on Vacuum Electronics, Beijing Vacuum Electronics Research Institute, Beijing 100015, China; youhunhaha123@163.com (A.L.); gaoshang19941210@163.com (S.G.); 18200123346@163.com (Z.W.); zyc913374137@163.com (Y.Z.); Zhangxiongzi@163.com (Z.Z.); fengjinjun@tsinghua.org.cn (J.F.)
[2] School of Electronics Engineering and Computer Science, Peking University, Beijing 100871, China
* Correspondence: zengxu1108@163.com (X.Z.); duchaohai@pku.edu.cn (C.D.)

Abstract: The gyrotron travelling wave tube (gyro-TWT) is an ideal high-power, broadband vacuum electron amplifier in millimeter and sub-millimeter wave bands. It can be applied as the source of the imaging radar to improve the resolution and operating range. To satisfy the requirements of the W-band high-resolution imaging radar, the design and the experimentation of the W-band broadband TE$_{02}$ mode gyro-TWT were carried out. In this paper, the designs of the key components of the vacuum tube are introduced, including the interaction area, electron optical system, and transmission system. The experimental results show that when the duty ratio is 1%, the output power is above 60 kW with a bandwidth of 8 GHz, and the saturated gain is above 32 dB. In addition, parasitic mode oscillations were observed in the experiment, which limited the increase in duty ratio and caused the measured gains to be much lower than the simulation results. For this phenomenon, the reasons and the suppression methods are under study.

Keywords: broadband; gyro-TWT; high-resolution imaging radar; TE$_{02}$ mode; W-band

Citation: Zeng, X.; Du, C.; Li, A.; Gao, S.; Wang, Z.; Zhang, Y.; Zi, Z.; Feng, J. Design and Preliminary Experiment of W-Band Broadband TE$_{02}$ Mode Gyro-TWT. *Electronics* **2021**, *10*, 1950. https://doi.org/10.3390/electronics10161950

Academic Editors: Mikhail Glyavin and Paolo Baccarelli

Received: 30 June 2021
Accepted: 10 August 2021
Published: 13 August 2021

Publisher's Note: MDPI stays neutral with regard to jurisdictional claims in published maps and institutional affiliations.

1. Introduction

The gyro-TWT is a vacuum electron amplifier based on the principle of the relativistic electron cyclotron maser. It can be used as the source to generate the high-power (kW levels) and broadband RF output in millimeter and sub-millimeter wave bands [1–5]. It is suitable as an important component of the transmitter, which can be applied to the radar, telecommunication, etc. Especially in W-band, the high-power, high-resolution imaging radar plays an important role in the applications of satellite imaging and deep space detection. A Haystack Ultra-wideband Satellite Imaging Radar (HUSIR) has been developed in the United States, in which a W-band gyro-TWT has been used as a part of the amplification link [6,7]. In order to achieve the detectability of an object with a diameter of 10 cm, the required operating bandwidth and minimum output power of the radar are 8 GHz and 100 kW, respectively. In the HUSIR, a W-band gyro-TWT with a bandwidth of 8 GHz and output power of 1 kW has been used to drive 16 gyrotwystrons to achieve the required capabilities. Therefore, under the precondition of sufficient bandwidth, increasing the output power is beneficial to the practical application of the gyro-TWT.

Reviewing the previous experiment results of the W-band gyro-TWT [8–13], all the designs adopt the fundamental harmonic of the TE$_{01}$ cylindrical waveguide mode as the operating mode, because of its high beam–wave interaction efficiency. At CPI (Communication & Power Industries), the designed tube has achieved an output power of 1.5 kW with a bandwidth of 8 GHz [8,9], which has been applied to the HUSIR. At UC Davis, the designed tube based on the heavily loaded and short copper stage interaction region achieved an output power of 140 kW, but the bandwidth was only 2 GHz [9–13]. At

IAP (Institute of Applied Physics, Russian Academy of Sciences), a W-band high-gain gyro-TWT based on the helical-waveguide was presented. The greatest advantage of this is that it greatly reduces the requirement of operating magnetic field intensity. The PIC simulations predict the maximum output power of 3 kW at about 95 GHz and a 1-kW-level bandwidth of 7.2 GHz when driving by a 100 mW input signal [14–18]. In China, research on W-band gyro-TWTs has been receiving more and more attention in recent years. At UESTC (University of Electronic Science and Technology of China), the periodic lossy circuit has been applied to the interaction region with an output power of 112 kW and a bandwidth of 3.8 GHz being achieved in the experiment [19,20]. At BVERI (Beijing Vacuum Electronics Research Institute), we also used the periodic lossy circuit as the interaction region of the tube, and obtained an output power of 110 kW and a bandwidth of 4 GHz [21].

However, an inner groove on the tube wall caused by the bombardment of the electrons had been observed in a previous experiment. The phenomenon indicates that a poor electron beam transmission seriously limits the output power of the tube. According to the results of theoretical calculations, the interaction cavity radius of the TE_{01} mode is only 2 mm, and the distance between the electron beam and the tube wall is less than 1 mm. It is very hard to avoid the bombardment of the electrons. Consequently, improvement of the electron beam transmission and the output power capacity is very difficult. To solve this problem, a fundamental harmonic of the TE_{02} cylindrical waveguide mode has been applied as the operating mode in our recent design. The interaction cavity radius of the TE_{02} mode is 3.7 mm, which greatly decreases the risk of the bombardment on the tube wall of the electrons; furthermore, the performance of the tube is guaranteed effectively.

The paper is organized as follows. The theoretical analysis of the beam–wave interaction is introduced in Section 2. The third section introduces the designs of the W-band TE_{02} mode gyro-TWT. The performance of the proposed tube is verified by experiment results and the problem of parasitic mode oscillation of the tube is analyzed in Section 4. Finally, the conclusion of the article and the overview of future work are presented in Section 5.

2. Theoretical Analysis

In the gyro-TWT, the amplification is achieved by using a cyclotron electron beam to interact with the transmitting microwave, which can be described by the following dispersion equations [22,23]:

$$\omega = k_z v_z + s\Omega \tag{1}$$

$$\omega^2 - k_z^2 c^2 - k_{mn}^2 c^2 = 0 \tag{2}$$

$$k_{mn} = \chi_{mn}/r_w \tag{3}$$

Here, ω is the operating angle frequency, k_z is the axial wave number, s is the cyclotron harmonic number, v_z is the axial velocity, and $\Omega = eB/(\gamma m_0)$ is the relativistic cyclotron frequency.

The interaction cavity radius can be determined by:

$$r_w = c\chi_{mn}/\omega \tag{4}$$

In the above, χ_{mn} is the nth root of the derivative with respect to x of $J_m(x)$, the mth-order Bessel function of the first kind. The χ_{mn} of the TE_{02} mode is 7.016, which is roughly two times larger than the χ_{mn} of the TE_{01} mode. The interaction cavity radius is increased significantly at the same operating frequency. However, the number of competing modes will also increase (shown in Figure 1). Figure 1 shows that the possible competing modes are TE_{22}, TE_{12}, TE_{01}, TE_{21} and TE_{11}; meanwhile, they will oscillate at frequencies of 85.5 GHz, 76.2 GHz, 72 GHz, 70.9 GHz, 69.2 GHz and 67 GHz, respectively. Therefore, the method of suppressing these competing modes is critical in the design.

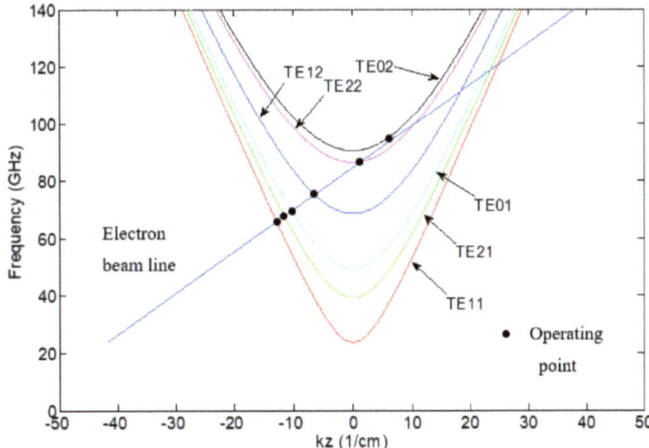

Figure 1. The dispersion diagram of the TE$_{02}$ mode and the possible oscillating modes (U = 70 kV, α = 1.2, B/B_g = 1.026, and r_w = 3.7 mm).

According to the observations from the previous experiment for the W-band TE$_{01}$ mode gyro-TWT, in order to suppress these competing modes, a periodic dielectric loaded circuit (shown in Figure 2) has been applied, in which the relative permittivity and the loss tangent of the dielectric, the periodicity, and the dimension of the circuit have been optimized for the highest efficiency.

Figure 2. The schematic of the periodic dielectric loaded circuit.

Meanwhile, the harmonic coupling coefficient that represents the interaction strength is defined as [22]:

$$H_{sm}\left(r_g, r_L\right) \;=\; J_{s \pm m}^2\left(x_{mn} r_g / r_w\right) J_s^{\prime 2}\left(x_{mn} r_L / r_w\right) \tag{5}$$

Here, r_L, r_g, and r_w are the Larmor radius, the guiding-center radius of the electrons, and the interaction cavity radius, respectively, with $\pm$ indicating co-rotating ($-$) and counter-rotating ($+$) modes, respectively. Figure 3 shows the dependence of the coupling coefficient on guiding-center radius for the TE$_{02}$ mode and the possible competing modes. In this figure, the coupling coefficient peaks at r_g / r_w = 0.26 for the TE$_{02}$ operating mode. Although this is smaller than the coupling coefficient of the TE$_{01}$ mode, the smaller guiding-center radius contributes to the increase in electron flow rate. In addition, the coupling coefficients of the other possible competing modes are very small; the value of r_g / r_w is 0.26, which indicates the oscillations of these modes have less influence on the performance of the tube.

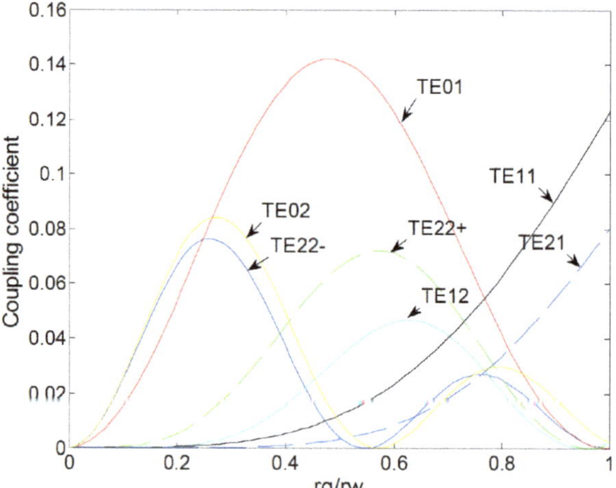

Figure 3. The dependence of the coupling coefficient on guiding-center radius for the TE_{02} mode and possible competing modes.

3. Designs and Simulation Results

In general, a gyro-TWT includes a magnetic injection gun, an input coupler, an interaction region, a collector, and an output window. The prototype of the W-band TE_{02} mode gyro-TWT is shown in Figure 4. In our design, a triode-type magnetic injection gun with a high voltage of 70 kV and a relatively low beam current of 9 A has been applied. The input coupler consists of a pillbox window and a Y-type mode convertor. The interaction region uses a periodic beryllium oxide ceramic-loaded circuit. The collector is cooled by water. The output window is formed by three pieces of sapphire discs. The operating parameters are listed in Table 1.

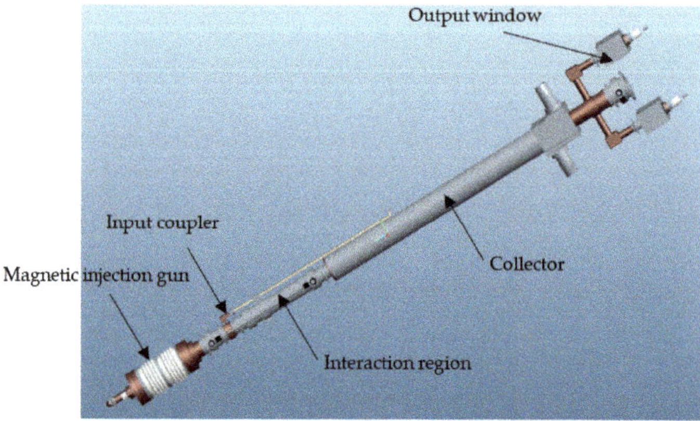

Figure 4. Schematic of the W-band TE_{02} mode gyro-TWT.

Table 1. Operating parameters of the W-band TE_{02} mode gyro-TWT.

Parameter	Value
Accelerating voltage	70 kV
Electron beam current	9 A
Velocity ratio	1.2
Operating mode	TE_{02}
Output mode	TE_{01}
Cyclotron harmonic	1
Magnetic field	3.4 T
Interaction cavity radius	3.7 mm
Guiding-center radius	0.96 mm
Relative permittivity of dielectric	12
Loss tangent of dielectric	0.225
Velocity spread	5%

3.1. Interaction Region

The interaction region is the most important component of the gyro-TWT, where the energy of the electron beam has been transferred to the RF wave when the synchronous condition is satisfied. According to the analysis in Section 2, the parasitic mode oscillations need to be suppressed. When the higher order mode is applied, the influence of these oscillations on the output power of the operating mode increases significantly. Figure 5 shows the process of the beam–wave interaction at a frequency of 94 GHz. By introducing the periodic dielectric loaded circuit, the parasitic mode oscillations are suppressed. In the smooth waveguide, the power of the input RF wave is amplified from 10 mW to 160 kW with an electron efficiency of 29%.

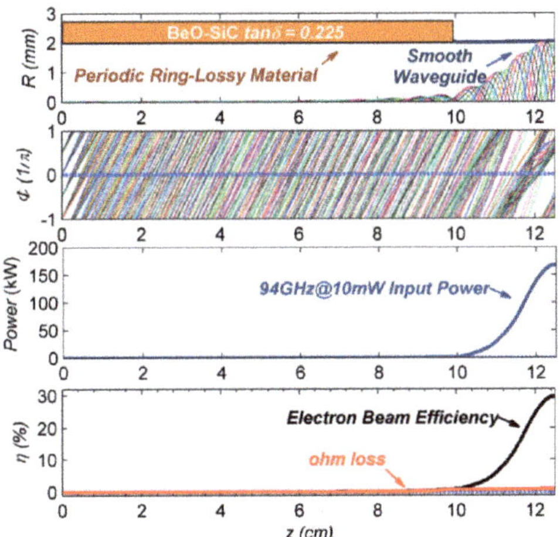

Figure 5. Process of the beam–wave interaction in the designed interaction region (f = 94 GHz, drive power is 10 mW, velocity spread is 5%).

Figure 6 shows the calculated instantaneous bandwidth of the W-band TE_{02} mode gyro-TWT. The results show that the 120 kW-level bandwidth is at 8.65 GHz, which satisfies the requirements of the application.

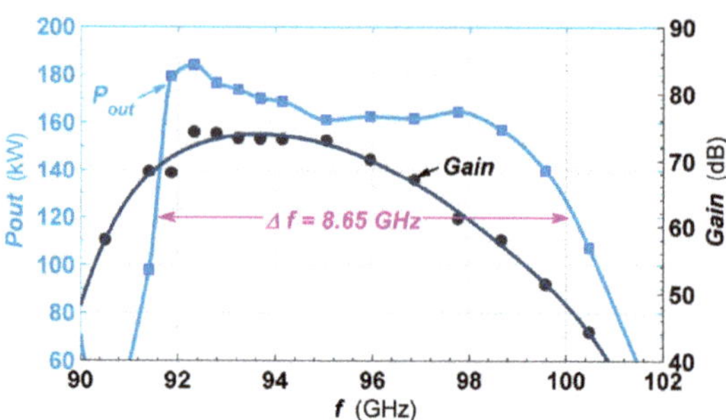

Figure 6. Calculated bandwidth of the W-band TE$_{02}$ mode gyro-TWT.

3.2. Electron Optical System

The electron optical system consists of: (1) a magnetic injection gun (MIG) for generating the desired rotating electron beam; (2) a collector, which is used to sort the residual electrons after interaction. This particular section deals with the design studies on the electron optical system, which are as follows.

3.2.1. Magnetic Injection Gun

A triode-type MIG consists of an emitter ring, a modulating anode and an accelerating anode. It is preferred for performance verification experiments by adjusting the voltage of the modulating anode to control the electron beam parameters. According to the beam–wave interaction calculation, the velocity ratio, guiding-center radius, and velocity spread are set as the design objectives of the MIG. By optimizing the structure dimensions and the potentials of the electrodes, the desired beam parameters were achieved, and carried out by Opera-3D code (seen in Figure 7 and Table 2).

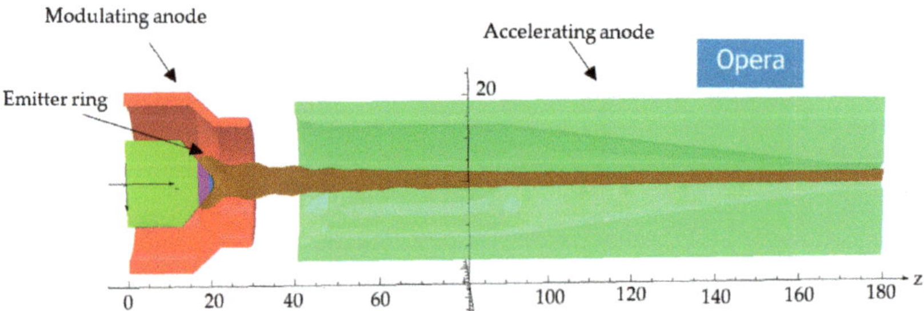

Figure 7. Electron trajectories of the optimized MIG.

Table 2. Beam properties of the optimized MIG.

Parameter	Value
Accelerating voltage	70 kV
Modulating voltage	30.7 kV
Electron beam current	9 A
Velocity ratio	1.2
Guiding-center radius	0.972 mm
Velocity spread	3.49%

3.2.2. Collector

In order to take out the heat generated by the dissipation power of the residual electrons in time, the collector is cooled by water. The optimization of the cooling structure was carried out by ANSYS. Under the parameters of water flow at 7 t/h and dissipation power density at 490 W/cm^2 (assuming all the electron energy transforms to heat and the duty ratio is 20%), the results of the temperature distribution are shown in Figure 8. The temperatures of the inner wall and outer surface are 136 °C and 94 °C, respectively, which are far below the outgassing temperature of the copper. According to the calculation, the average power capacity of the designed collector is above 130 kW.

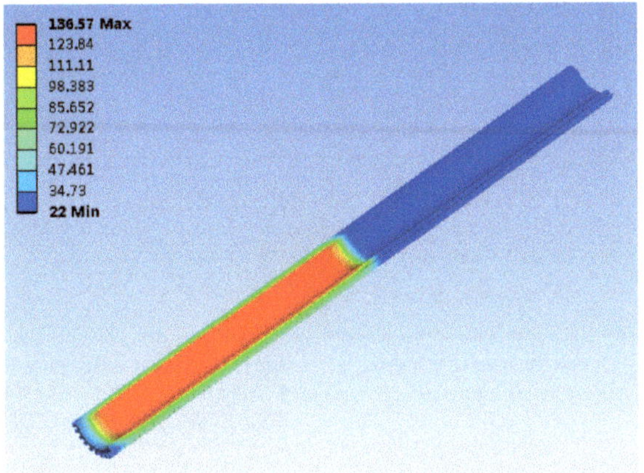

Figure 8. Temperature distribution of the optimized collector.

3.3. Transmission System

The transmission system includes: (1) an input coupler, which is used to transform the rectangular waveguide TE$_{10}$ mode to the cylindrical waveguide TE$_{02}$ mode; (2) an output window, which is used to maintain a high-vacuum environment in the tube and transmit the cylindrical waveguide TE$_{01}$ mode to the outside of the tube with low reflection. The bandwidth and reflection coefficient are the most important properties of the transmission system.

3.3.1. Input Coupler

In our design, the input coupler consists of a pillbox window and a mode converter. The pillbox window is made of CVD diamond, which has a perfect matching performance in a W band. In the mode converter, a Y-type structure has been applied [24–26]. A standard rectangular waveguide has been divided into eight ports to connect with a cylindrical waveguide; these ports are uniformly distributed along the circumference. The simulation results show that the TE$_{02}$ mode can be excited directly from the TE$_{10}$ mode.

Figure 9 shows the calculated VSWR of the input coupler. From 90 GHz to 102 GHz, the VSWRs are between 1.2 and 2.1, which meet the requirements of the broadband power drive. The fluctuation of the VSWR is caused by the reflections of the eight ports.

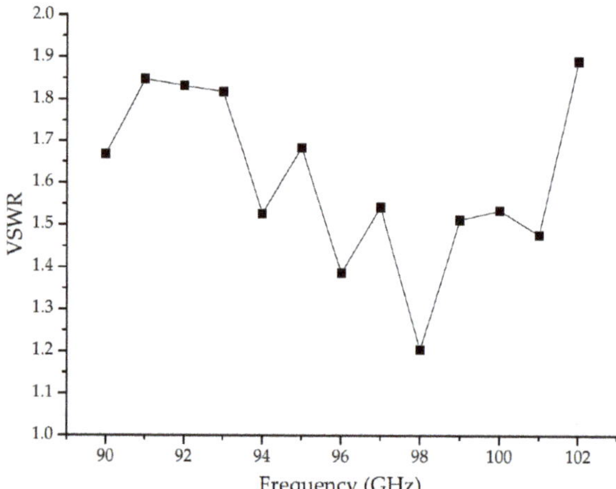

Figure 9. Calculated VSWR of the input coupler.

3.3.2. Output Window

To achieve broadband operation, a triple-sapphire-disc configuration has been applied on the output window [24]. Based on the view of the resonant window, the thickness of the disc is about $N\bullet\lambda_d/2$, where N is equal to 1,2,3 ... , and λ_d is the wavelength of sapphire. By optimizing the thicknesses of the discs and the distances between the discs, the lowest reflection coefficient was achieved.

The diameter of the output window is 32 mm. After optimization, the thicknesses of the three discs are 0.4 mm, 0.8 mm, and 0.4 mm, respectively. The distances between the discs are 0.51 mm. The relative dielectric constant and the loss tangent of the sapphire are 9.5 and 2×10^{-4}, respectively. Figure 10 shows the calculated VSWR of the output window. In 91–100 GHz, the VSWRs are less than 1.1, which indicate that the advantageous property of low reflection is obtained.

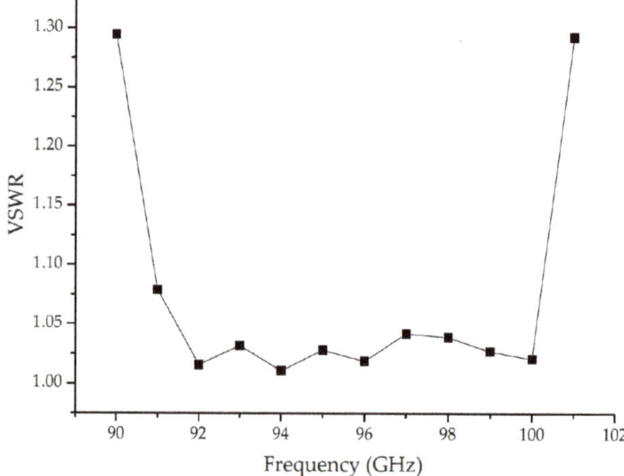

Figure 10. Calculated VSWR of the output window.

4. Experiment Results and Discussion

According to the design scheme of the W-band TE_{02} mode gyro-TWT, fabrication of the components was carried out. Before assembling the tube, the critical components, such as the input coupler and output window, were tested by the VNA (Vector Network Analyzer), as seen in Figure 11. The cold test results are shown in Figure 12. In 92–100 GHz, the measured VSWRs of the input coupler are less than 2, and the measured VSWRs of the output window are less than 1.1. In consideration of the influences of the machining errors, the above results agree well with the simulation results.

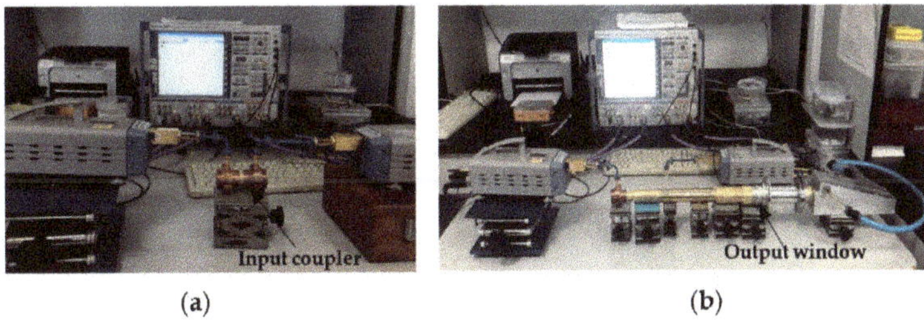

Figure 11. Cold test of the critical components of the W-band gyro-TWT by VNA (**a**) Input coupler; (**b**) Output window.

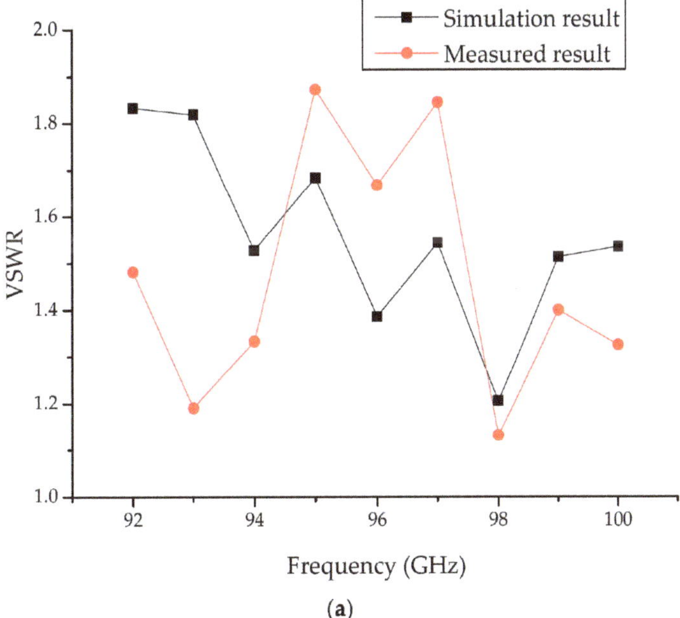

Figure 12. *Cont.*

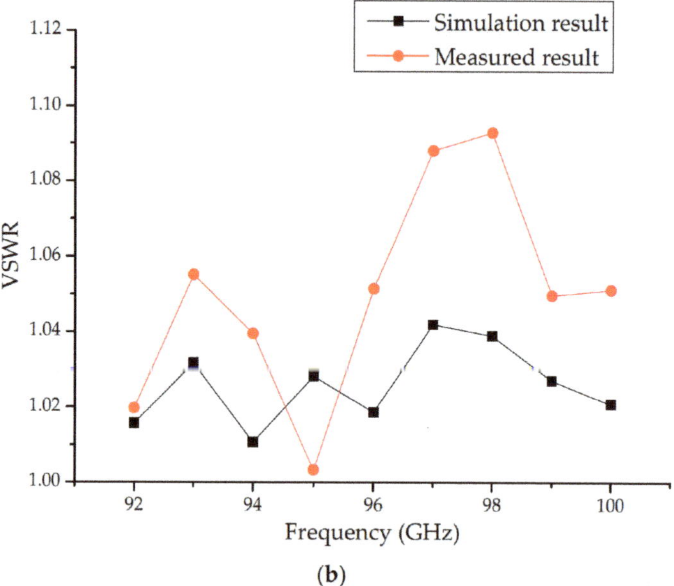

(b)

Figure 12. Cold test results of the critical components of the W-band gyro-TWT. (**a**) Input coupler; (**b**) Output window.

The prototype of the W-band TE_{02} mode gyro-TWT is shown in Figure 13. The length and maximum diameter of the tube are 1400 mm and 85 mm, respectively.

Figure 13. Prototype of the W-band TE_{02} mode gyro-TWT.

The hot test system diagram of the tube is shown in Figure 14. A driving signal which is generated by a W-band signal generator is amplified by a W-band solid state amplifier and a W-band TWT. The maximum output power of the W-band TWT is 150 W. The W-band gyro-TWT was installed on a superconducting magnet with a maximum magnetic field intensity of 4 T. The voltage of the W-band gyro-TWT is provided by a high-voltage power supply. The output power of the tube is measured by a water load and a calorimeter. The waveforms of the beam voltage, beam current, driving signal, and output signal are measured by a four-channel oscilloscope. A spectrum analyzer is used to detect the operating mode and parasitic modes.

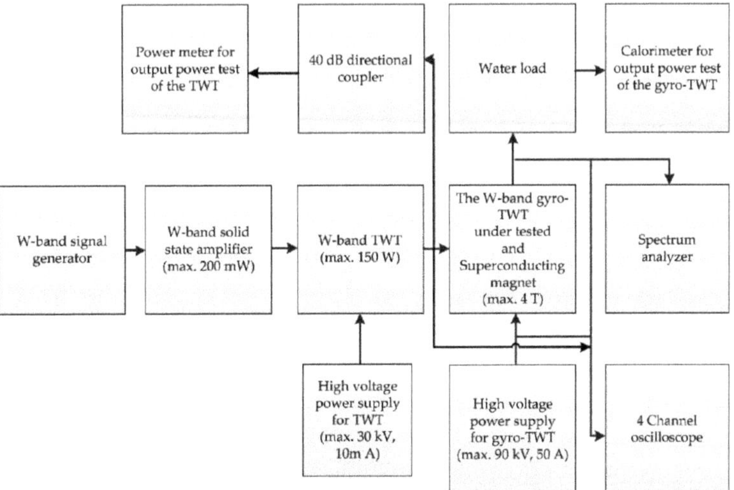

Figure 14. Hot test system diagram of the W-band TE_{02} mode gyro-TWT.

First, experimental verification was carried out. In the duty ratio of 1%, the accelerating voltage was 70.2 kV, the electron beam current was 8.5 A, and the maximum magnetic field was 3.46 T. The measured output power and gain are shown in Figure 15. It can be seen that from 91 to 99 GHz, the output power is above 60 kW, and the gain is above 32 dB. The property of broadband power amplification of the W-band TE_{02} mode gyro-TWT was verified.

However, the output power and the gain are much lower than the simulation results, due to the oscillations in the tube. The frequency spectrums of the output signal were detected in the experiment by a spectrum analyzer, as shown in Figure 16. The amplification of the operating mode and the oscillations were observed. According to the dispersion diagram in Figure 1, the oscillations correspond to the absolute instability of the TE_{02} mode at 88 GHz and the backward wave oscillation of the TE_{22} mode at 85 GHz, respectively. These oscillations disturb the movements of the electron beams and further lead to the increase in velocity spread. Therefore, the beam–wave interaction efficiency decreases rapidly. In addition, they have a significant influence on the operating stability of the tube, which restricts the increase in duty ratio. In subsequent improvements, a heavier dielectric loss will be applied in the interaction region to suppress the mode oscillation.

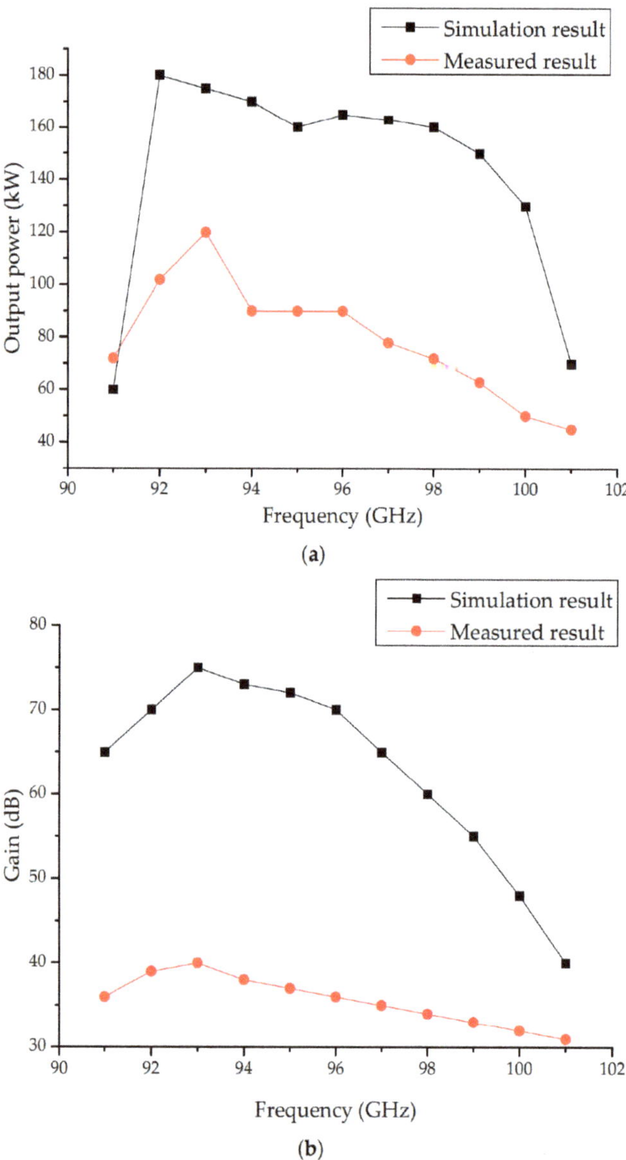

(**a**)

(**b**)

Figure 15. Measured output power (**a**) and gain (**b**) of the W-band TE$_{02}$ mode gyro-TWT.

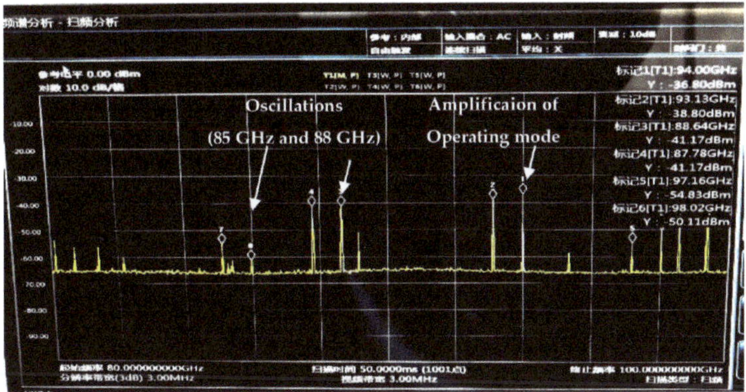

Figure 16. Measured frequency spectrums of the W-band TE_{02} mode gyro-TWT.

5. Conclusions

In this paper, the design of a W-band TE_{02} mode gyro-TWT is presented. The simulation results indicate that the performances of the tube meet the requirements of the output power greater than 100 kW and the bandwidth greater than 8 GHz. Fabrication of the components was carried out. The cold test results of the transmission system agree well with the simulation results, which verify that the transmission system has the advantageous properties of broadband and loss reflection. The prototype of the W-band TE_{02} mode gyro-TWT was assembled. In the preliminary experiment, a bandwidth of 8 GHz was achieved, and the output powers were above 60 kW at 70.2 kV/8.5 A. The feasibility of using the TE_{02} mode as the operating mode to achieve broadband amplification in the W-band is verified. However, the measured results are far below the simulation's results, which are caused by the oscillations in the tube. The mode oscillations are observed, which causes the beam–wave interaction efficiency to decrease rapidly and the duty ratio to be restricted to 1%. An improvement of applying a heavier dielectric loss to suppress the oscillations is in progress. The reasons for this and the suppression methods are also under study.

Author Contributions: Conceptualization, X.Z., C.D. and J.F.; methodology, C.D.; validation, X.Z., C.D., Y.Z.; formal analysis, A.L., S.G., Z.W.; investigation, X.Z., C.D., Z.W. and Z.Z.; resources, J.F.; data curation, X.Z. and Z.W.; writing—original draft preparation, X.Z. and Z.W.; writing—review and editing, X.Z. and J.F.; visualization, Y.Z.; supervision, J.F.; project administration, X.Z.; funding acquisition, X.Z. All authors have read and agreed to the published version of the manuscript.

Funding: This research received no external funding.

Data Availability Statement: All data included in this study are available upon request by contacting with the corresponding author.

References

1. Chu, K.R. The electron cyclotron maser. *Rev. Modern Phys.* **2004**, *76*, 489–540. [CrossRef]
2. Kartikeyan, M.V.; Borie, E.; Thumm, M. *Gyrotrons: High Power Microwave and Millimeter Wave Technology*; Springer: Berlin/Heidelberg, Germany; New York, NY, USA, 2003; pp. 23–25.
3. Booske, J.H.; Dobbs, R.J.; Joye, C.D.; Kory, C.L.; Neil, G.R.; Park, G.S.; Park, J.; Temkin, R.J. Vacuum electronic high power terahertz sources. *IEEE Trans. Terahertz Sci. Technol.* **2011**, *1*, 54–75. [CrossRef]
4. Nusinovich, G.S.; Thumm, M.K.A.; Petelin, M.I. The gyrotron at 50: Historical overview. *J. Infr. Milli. Terahz. Waves* **2014**, *35*, 325–381. [CrossRef]

5. Jiang, W.; Lu, C.; Liu, Y.; Wang, J.; Liu, G.; Luo, Y. A Novel Multi-beam Magnetron Injection Gun for W-band Gyro-TWT. In Proceedings of the 22nd IEEE International Vacuum Electronics Conference (IVEC 2021), Virtual, 28–30 April 2021; pp. 428–429.
6. Kempkes, M.A.; Hawkey, T.J.; Gaudreau, M.P.J.; Phillips, R.A. W-Band Transmitter Upgrade for the Haystack Ultra-wideband Satellite Imaging Radar (HUSIR). In Proceedings of the IEEE Vacuum Electronics Conference, Scottsdale, AZ, USA, 8–11 June 2006; pp. 551–552.
7. MacDonald, M.E.; Anderson, J.P.; Lee, R.K. The HUSIR W-Band Transmitter. *Linc. Lab J.* **2014**, *21*, 106–114.
8. Blank, M.; Borchard, P.; Cauffman, S.; Felch, K. Demonstration of a broadband W-band gyro-TWT amplifier. In Proceedings of the International Vacuum Electronics Conference and International Vacuum Electron Sources (IVEC/IVESC 2006), Monterey, CA, USA, 25–27 April 2006; pp. 459–460.
9. Blank, M.; Borchard, P.; Cauffman, S.; Felch, K. Design and Demonstration of W-band Gyrotron Amplifiers for Radar Applications. In Proceedings of the 2007 Joint 32nd International Conference on Infrared and Millimeter Waves and 15th International Conference on Terahertz Electronics 2007, Cardiff, UK, 2–9 September 2007; pp. 358–361.
10. Song, H.H.; Barnett, L.R.; McDermott, D.B.; Hirata, Y.; Hsu, H.I.; Marandos, P.S.; Lee, J.S.; Chang, T.H.; Chu, K.R.; Luhmann, N.C., Jr. W-band heavily loaded TE01 gyrotron traveling wave amplifier. In Proceedings of the 4th IEEE International Vacuum Electronics Conference (IVEC 2003), Seoul, Korea, 28–30 May 2003; pp. 348–349.
11. Song, H.H.; McDermott, D.B.; Hirata, Y.; Barnett, L.R.; Domier, C.W.; Hsu, H.L.; Chang, T.H.; Tsai, W.C.; Chu, K.R.; Luhmann, N.C., Jr. Theory and experiment of a 95 GHz gyrotron traveling wave amplifier. In Proceedings of the Conference Digest 28th International Conference on Infrared and Millimeter Waves 2003, Otsu, Japan, 29 September–2 October 2003; pp. 205–206.
12. Song, H.H.; McDermott, D.B.; Hirata, Y.; Barnett, L.R.; Domier, C.W.; Hsu, H.L.; Chang, T.H.; Tsai, W.C.; Chu, K.R.; Luhmann, N.C., Jr. Theory and experiment of a 94 GHz gyrotron traveling-wave amplifier. *Phys. Plasmas* **2004**, *11*, 2935–2941. [CrossRef]
13. Barnett, L.R.; Luhmann, N.C., Jr.; Chiu, C.C.; Chu, K.R.; Yan, Y.C. Advances in W-Band TE01 gyro-TWT amplifier design. In Proceedings of the 8th IEEE International Vacuum Electronics Conference (IVEC 2007) 2007, Kitakyushu, Japan, 15–17 May 2007; p. 233.
14. Sergey Samsonov, V.; Alexander Bogdashov, A.; Gregory Denisov, G.; Igor Gachev, G.; Mishakin, S.V. Cascade of Two W-Band Helical-Waveguide Gyro-TWTs With High Gain and Output Power: Concept and Modeling. *IEEE Trans. Electron Devices* **2017**, *64*, 1297–1301. [CrossRef]
15. Naum Ginzburg, S.; Gregory Denisov, G.; Michael Vilkov, N.; Zotova, I.V.; Alexander Sergeev, S.; Roman Rozental, M.; Sergey Samsonov, V.; Sergey Mishakin, V.; Marek, A.; Jelonnek, J. Ultrawideband Millimeter-Wave Oscillators Based on Two Coupled Gyro-TWTs With Helical Waveguide. *IEEE Trans. Electron Devices* **2018**, *65*, 2334–2338. [CrossRef]
16. Samsonov, S.V.; Bogdashov, A.A.; Denisov, G.G.; Gachev, I.G. Experiments on W-band High-Gain Helical-Waveguide Gyro-TWT. In Proceedings of the 20th IEEE International Vacuum Electronics Conference (IVEC 2019), Busan, Korea, 29 April–1 May 2019; pp. 348–349.
17. Samsonov, S.V.; Denisov, G.G.; Gachev, I.G.; Bogdashov, A.A. CW Operation of a W-Band High-Gain Helical-Waveguide Gyrotron Traveling-Wave Tube. *IEEE Trans. Electron Devices* **2020**, *41*, 773–776. [CrossRef]
18. Samsonov, S.V.; Leshcheva, K.A.; Manuilov, V.N. Multitube Helical-Waveguide Gyrotron Traveling-Wave Amplifier: Device Concept and Electron-Optical System Modeling. *IEEE Trans. Electron Devices* **2020**, *99*, 1–6. [CrossRef]
19. Yan, R.; Tang, Y.; Luo, Y. Design and experimental study of a high-gain W-band gyro-TWT with nonuniform periodic dielectric loaded waveguide. *IEEE Trans. Electron Devices* **2014**, *61*, 2564–2569.
20. Wang, J.; Yao, Y.; Tian, Q.; Li, H.; Liu, G.; Luo, Y. Broadband High-Efficiency Input Coupler With Mode Selectivity for a W-Band Confocal Gyro-TWA. *IEEE Trans. Microw. Theory Tech.* **2018**, *66*, 1895–1901. [CrossRef]
21. Wang, E.F.; Xi, A.; Zeng, X. Preliminary experiment research on the W-band gyrotron traveling wave tube. In Proceedings of the16th IEEE International Vacuum Electronics Conference (IVEC 2015), Beijing, China, 27–29 April 2015; pp. 98–99.
22. Chu, K.R.; Lin, A.T. Gain and bandwidth of the gyro-TWT and CARM amplifiers. *IEEE Trans. Plasma Sci.* **1988**, *16*, 90–104. [CrossRef]
23. Chu, K.R.; Drobot, A.T.; Szu, H.H.; Sprangle, P. Theory and Simulation of the Gyrotron Traveling Wave Amplifier Operating at Cyclotron Harmonics. *IEEE Trans. Microw. Theory Tech.* **1980**, *28*, 313–317.
24. Zeng, X.; Wang, E.; Cai, J.; Feng, J. Study on Ka-band broadband mode converter for gyro-TWT. In Proceedings of the 19nd IEEE International Vacuum Electronics Conference (IVEC 2018), Monterey, CA, USA, 24–26 April 2018; pp. 235–236.
25. Fang, C.; Liu, G.; Rao, W.; Wang, Y.; Wang, S.; Wang, J.; Jiang, W.; Wang, L.; Luo, Y.; Shu, G. A Novel TE01 Input Coupler for a W-band Gyrotron Traveling-Wave Tube. In Proceedings of the 20nd IEEE International Vacuum Electronics Conference (IVEC 2019), Busan, Korea, 29 April–1 May 2019; pp. 331–332.
26. Du, C.H.; Chang, T.H.; Liu, P.K.; Huang, Y.C.; Jiang, P.X.; Xu, S.X.; Geng, Z.H.; Hao, B.L.; Xiao, L.; Liu, G.F. Design of a W-band Gyro-TWT Amplifier With a Lossy Ceramic-Loaded Circuit. *IEEE Trans. Electron Devices* **2013**, *60*, 2388–2394. [CrossRef]

 *electronics*

Article

Experimental Investigation into the Optimum Position of a Ring Reflector for an Axial Virtual Cathode Oscillator

Se-Hoon Kim, Chang-Jin Lee, Wan-Il Kim and Kwang-Cheol Ko *

Department of Electrical Engineering, Hanyang University, Seoul 04763, Korea;
d.sehoon.kim@gmail.com (S.-H.K.); jincalibur@naver.com (C.-J.L.); wanilgg@naver.com (W.-I.K.)
* Correspondence: kwang@hanyang.ac.kr

Abstract: A ring reflector was experimentally investigated using an axial virtual cathode oscillator (vircator). The ring reflector was installed behind the mesh anode of the axial vircator to enhance the microwave power output by forming a resonant cavity and increasing the electron beam to microwave energy conversion efficiency. The optimum position of the ring reflector is analyzed through simulations and experiments by varying the anode to reflector distance from 6 mm to 24 mm in 3 mm steps. PIC simulations show that the ring reflector enhances the microwave power of the axial vircator up to 220%. Experiments show that the microwave power from the axial vircator without the ring reflector is 11.22 MW. The maximum average peak microwave power of the axial vircator with the ring reflector is 25.82 MW when the anode to ring reflector distance is 18 mm. From the simulations and experiments, it can be seen that the ring reflector yields decaying enhancement that is inversely proportional to the anode to ring reflector distance and there is no noticeable microwave enhancement after 24 mm. The frequency range attained from the simulations and experiments is 5.8 to 6.7 GHz and 5.16 to 5.8 GHz, respectively. The difference between the simulation and experimental results is due to the error in the anode to cathode gap distance. Although the frequency is slightly changed, the ring reflector seems to have no influence on the frequency of the generated microwave.

Keywords: high-power microwave source; HPM source; virtual cathode oscillator; vircator; ring reflector

check for updates

Citation: Kim, S.-H.; Lee, C.-J.; Kim, W.-I.; Ko, K.-C. Experimental Investigation into the Optimum Position of a Ring Reflector for an Axial Virtual Cathode Oscillator. *Electronics* **2021**, *10*, 1878. https://doi.org/10.3390/electronics10161878

Academic Editor: Noel Rodriguez

Received: 30 June 2021
Accepted: 3 August 2021
Published: 5 August 2021

Publisher's Note: MDPI stays neutral with regard to jurisdictional claims in published maps and institutional affiliations.

1. Introduction

As high-power microwave (HPM) devices have shown potential in various industries, many high-power electron devices have been studied and developed in many applications [1]. The HPM system utilizes pulsed power technologies and electron devices to generate microwaves from a few MW to a few GW. The HPM system is composed of a prime power source, a pulsed power system, and an HPM source. The virtual cathode oscillator (vircator) is one of the high-power electron devices used for the HPM source [2]. The vircator has low intrinsic impedance and is suitable for being driven using a low-impedance pulsed power source [3–6]. The vircator has two microwave-generating mechanisms. One is the reciprocating motion of electrons between the real cathode and the virtual cathode formed behind the anode, and the other is an oscillation of the virtual cathode. The vircator has been studied widely due to its structural simplicity and its ability to be tuned easily. The structural simplicity allows for high-voltage operation and the tuning characteristics allows for simple frequency modulation. However, the vircator has relatively low efficiency (typically below 5%) compared to other high-power electron devices. Many researchers have studied various types of vircators such as axial vircators, reflex triode vircators, and coaxial vircators to modulate the output microwave and increase its efficiency [7–14].

Researchers installed reflectors in the vircator to improve its efficiency and enhance its output power. Various types of reflectors were studied using particle-in-cell (PIC) simulations and experiments [15–20]. Among these reflectors, a ring-type reflector is reported to increase the electron beam to microwave power conversion efficiency by forming a resonant cavity when used in the coaxial vircator [15].

In this paper, an axial vircator with and without a ring-type reflector is experimentally investigated. The microwave power enhancement and the shift in the frequency spectrum due to the installation of the ring reflector have been analyzed through PIC simulations and experiments. A ring reflector is placed behind the mesh anode in the vircator chamber. To find the optimum anode to ring reflector distance, the ring reflector distance is changed from 6 mm to 24 mm in 3 mm steps. The axial vircator is driven using a 10-stage PFN-Marx generator. The influence of the ring reflector is analyzed by looking at the microwave power and the frequency for different anode to ring reflector distances. The overall system description is introduced in Section 2. The simulation and experimental results are analyzed in Section 3 and summarized in Section 4, consequently.

2. System Description

Figure 1 shows the schematic diagram of the experimental vircator system. The experimental system is composed of a pulsed power system, an HPM device, and a measurement system. The pulsed power system compresses the charged voltage and applies the high-voltage pulse to the load. The HPM device generates a high-power microwave using the supplied high-voltage pulse from the pulsed power source. An axial vircator is used as an HPM source in this experiment. As a measurement system, a receiving antenna is installed apart from the high-power microwave device to measure and analyze the characteristics of the pulsed power system and the high-power microwave device.

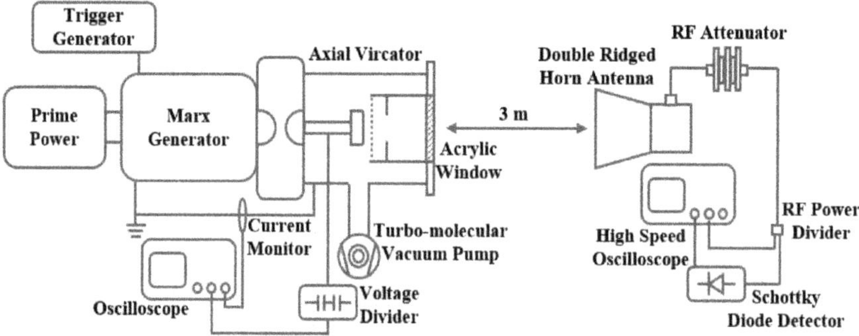

Figure 1. Schematic diagram of the experimental system.

2.1. PFN-Marx Generator

A 10-stage pulse-forming network (PFN)-Marx generator is selected to drive the HPM device. The PFN-Marx generator is suitable for driving low impedance HPM sources. The PFN consists of a consecutive array of L-C ladder circuit generating rectangular voltage pulses. When the PFNs are erected, the PFN-Marx generator produces a rectangular pulse with the pulse duration of the PFN which is multiplied by the charged voltage on the PFN and the number of stages [21–23]. By selecting proper inductance and capacitance values, the impedance of the PFN-Marx generator and that of the axial vircator can be matched without using additional impedance-matching devices, such as a tapered transmission line. The PFNs are constructed using high-voltage capacitors and copper strip inductors. Two PFN arrays with a characteristic impedance of 6 Ω are connected in parallel to reduce the characteristic impedance of the PFN module to 3 Ω. The resulting characteristic impedance of the PFN-Marx generator is 30 Ω. The design parameters of the 10-stage PFN-Marx generator are shown in Table 1. The capacitance and the inductance described in Table 1 is the composite value of two PFN arrays. A schematic circuit of the 10-stage PFN-Marx generator is shown in Figure 2. The circuit shown in Figure 2 is also used in the circuit simulation of the PFN-Marx generator.

Table 1. Design parameters of the 10-stage PFN-Marx generator.

Parameter	Value	Parameter	Value
Capacitance	4.17 nF	Inductance	37.5 nH
PFN stage	6	Marx stage	10
Charging voltage	−30 kV	Erected voltage	−300 kV
Pulse width	150 ns	Characteristic impedance	30 Ω

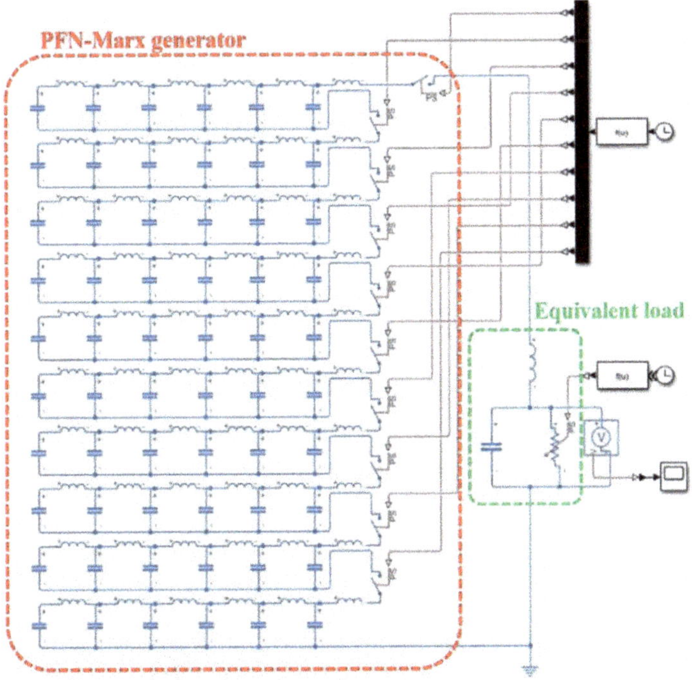

Figure 2. Schematic circuit of the 10-stage PFN-Marx generator.

Figure 3 shows the typical output waveform of the PFN-Marx generator of the simulations (a) and experiments (b). In the circuit simulation, the PFN-Marx generator is charged with a positive voltage. Alternatively, because the vircator chamber is grounded, the PFN-Marx generator is negatively charged in the experiments. The plateau voltage level of the simulation and experiments is approximately the same. The plateau voltage of −150 kV is used in the PIC simulations.

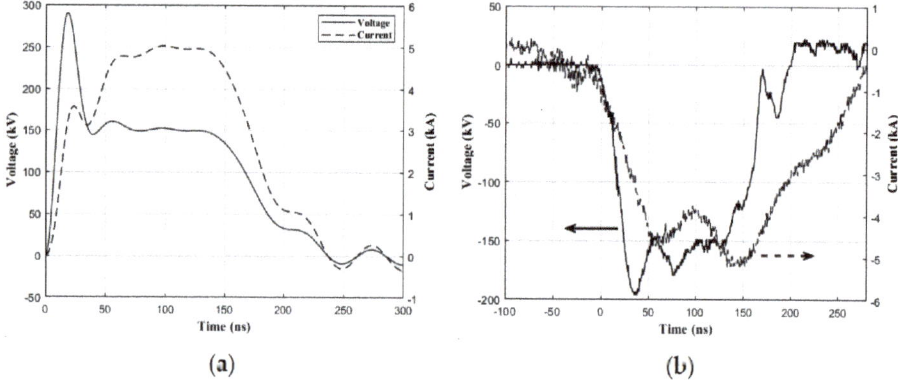

Figure 3. Typical output waveforms of the diode voltage and the diode current: (**a**) simulation results and (**b**) experimental results.

2.2. Axial Vircator

A vircator is a microwave source classified as a space charge device. It is one of various high-power microwave devices. Among various vircators, the axial vircator extracts the microwave along its axis. The axial vircator used in experiments is housed in a stainless-steel chamber (300 mm in diameter and 400 mm in length). A drift tube (200 mm in diameter) is installed in the chamber to attach the ring reflector and the anode assemblies. The stainless-steel chamber is evacuated to a pressure below 3×10^{-5} torr using a turbomolecular vacuum pump. The cathode holder and the back plate of the chamber are machined using poly-ether-ether ketone (PEEK) to prevent electrical breakdown between the voltage feeder and the vircator chamber. A stainless-steel mesh anode with geometric transparency of 70% and a graphite cathode are used as the vircator diode. The diameters of the cathode and the mesh anode are 70 mm and 200 mm, respectively. In the experiments, the anode to cathode gap (A-K gap) distance is fixed at 6 mm. The internal structure of the axial vircator is shown in Figure 4.

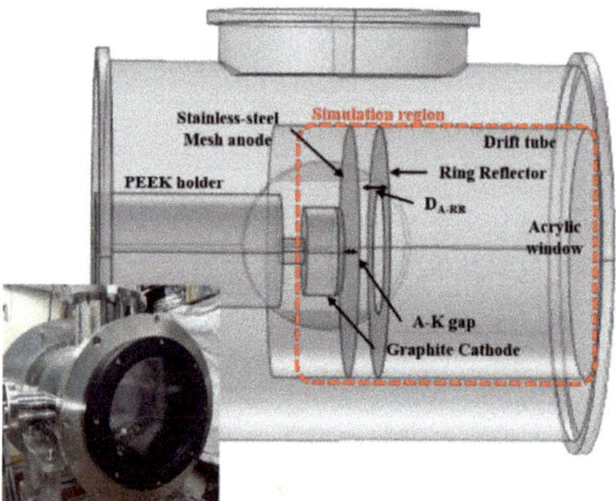

Figure 4. Inner structure of the axial vircator.

2.3. Measurement Equipment

To analyze the characteristics of the pulsed power system, the diode voltage and the current are measured using a capacitive voltage divider and a current monitor, respectively. A Pearson coil is used as the current monitor. Both the capacitive voltage divider and the Pearson coil are installed in the voltage feedthrough. The diode voltage and the current are recorded using an oscilloscope (DPO 3054).

To analyze the generated microwave power, a double ridged horn antenna is placed 3 m away from the vircator window. The vircator to antenna distance is selected to accord with the far-field criteria. A planar-doped barrier diode detector (8474B) is used to measure the received microwave power. The diode detector converts microwave power into voltage. The power-voltage conversion ratio of the detector is used in calculating the power from the vircator. The frequency of the microwave power is analyzed using a fast Fourier transform (FFT). The microwave signal and diode detector output are recorded using a high-speed oscilloscope (MSO 71604C). To protect the high-speed oscilloscope and the diode detector, a −30 dB attenuator is installed after the antenna. Considering the attenuation at the RF cable from the antenna to the oscilloscope and the insertion loss at the power divider, the total attenuation at the RF measuring system is −51 dB.

The microwave power at the axial vircator is calculated using Friis's equation. Friis's equation is given as:

$$P_r = \frac{P_r}{G_t G_r}\left(\frac{4\pi D}{\lambda}\right)^2 \tag{1}$$

Here, P_t is the power at the transmitting antenna, P_r is the power at the receiving antenna, G_t is the transmitting antenna gain, G_r is the receiving antenna gain, D is the distance between the transmitting antenna and the receiving antenna, and λ is the wavelength of the microwave signal [24]. In the RF measuring system, the gain of the transmitting antenna G_t and the gain of the double-ridged horn antenna G_r are 18.5 dBi and 12.82 dBi, respectively.

3. PIC Simulation and Experimental Results

The axial vircator with a ring reflector is investigated through PIC simulations and experiments. The ring reflector with a thickness of 1 mm is installed behind the mesh anode. The outer and inner diameter of the ring reflector is 200 mm and 140 mm, respectively. To optimize the cavity volume formed due to the ring reflector, the simulations and the experiments are conducted by varying the distance between the mesh anode and the reflector from 6 mm to 24 mm in 3 mm steps.

3.1. PIC Simulation

The axial vircator with and without the ring reflector is analyzed before experiments by using an FDTD-PIC simulation (CST particle studio). The simulation region is marked in Figure 4. Because the drift tube is the main operating region, the rest region except the drift tube region is excluded in the simulation model. The diameter and the length of the simulation space are 200 mm and 300 mm, respectively. The surface of the cathode and the anode are placed at z = 20 mm and z = 26 mm, respectively. The threshold voltage for electron emission is set to 100 kV/m. The mesh anode is modeled as a thin sheet with a transparency of 70%. A ramp-shaped voltage pulse with a pulse width of 25 ns and a plateau voltage of −150 kV is used as a vircator operating voltage waveform. Figure 5 shows the phase diagrams of the axial vircator. When the anode to ring reflector distance is between 12 mm and 21 mm, the discontinuous momentum points are distinctly shown around the reflector position. The simulation results showing the microwave power from the axial vircator are depicted in Figure 6. The microwave power is normalized based on that of the axial vircator without the ring reflector. According to the simulations, it can be seen that the ring reflector enhances the microwave power by up to 220%. However, the power enhancement ceased when the anode to ring reflector distance increased to 24 mm. FFT results are shown in Figure 7. The dominant frequencies are around 5.8 GHz when the ring reflector is not used or when the anode to ring reflector distance is larger than 18 mm. However, the dominant frequencies are above 6 GHz when the anode to ring reflector distance is between 6 mm and 15 mm. The reciprocating motion of electrons between the cathode and the virtual cathode is one of the principles of microwave generation in the vircator. According to the simulations, the ring reflector seems to attract the virtual cathode and lengthen the reciprocating distance when the anode to ring reflector distance is between 6 mm and 15 mm. Because the reciprocating frequency is inversely proportional to the reciprocating distance, the frequency shift is caused due to the lengthened reciprocating distance [24].

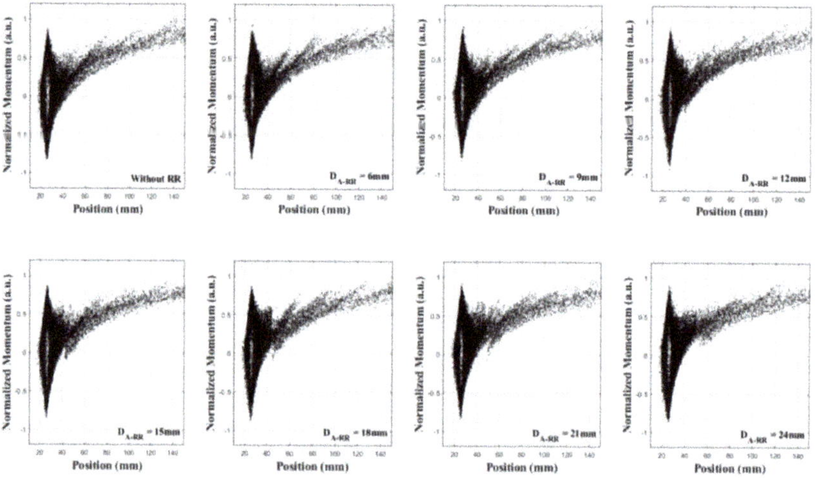

Figure 5. Phase space diagram of the axial vircator with and without the ring reflector.

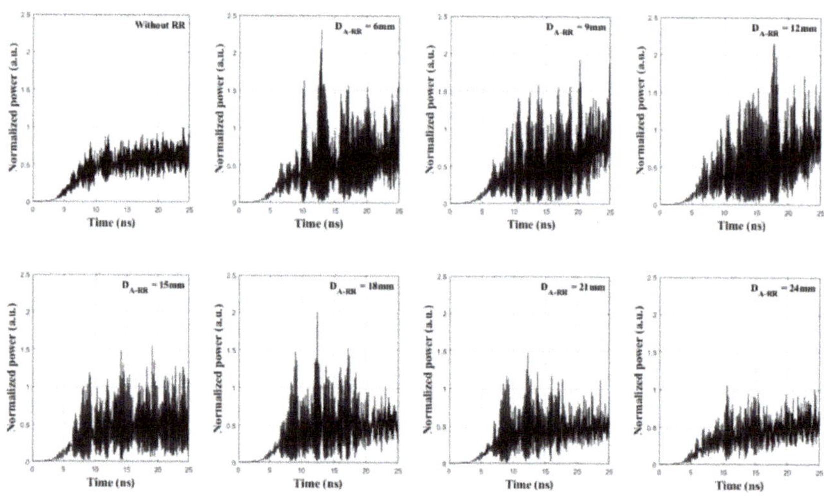

Figure 6. Normalized microwave power of the axial vircator with and without the ring reflector.

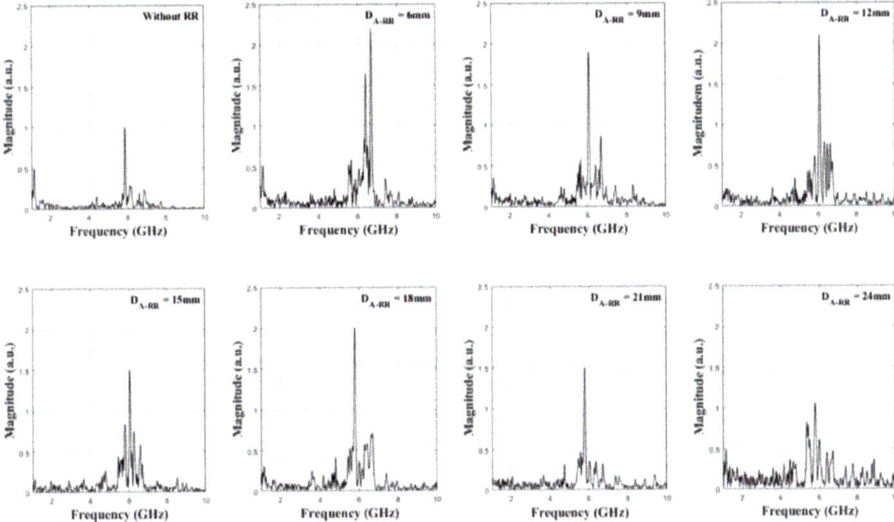

Figure 7. Frequency spectrum of the axial vircator with and without the ring reflector.

3.2. Experimental Results

The axial vircator is used to investigate and verify the enhancement of the microwave power caused by using the ring reflector. Figure 3b shows the typical voltage and current waveforms when $D_{A\text{-}RR}$ is 12 mm. The measured peak and plateau diode voltages were -196 ± 5 kV and -150 ± 7 kV, respectively. The rise time and the pulse width of the voltage pulse are 25 ns and 200 ns, respectively. The peak diode current is -5.6 ± 0.7 kA. According to the experiments, the presence and the position of the ring reflector have no noticeable influence on the diode voltage or the diode current. The output microwave from the axial vircator is calculated using the RF diode detector output voltage and the dominant frequency attained from the FFT results. The typical diode detector output waveforms when the axial vircator is driven without the ring reflector and with the ring reflector ($D_{A\text{-}RR}$ is 12 mm) are shown in Figure 8. Although the pulse width of the input voltage is 200 ns, the pulse width of the RF diode detector output is 140 ns. According to the input voltage and the RF detector output waveforms, the microwave is considered to be generated at the plateau part of the input voltage. Figure 9 shows the typical frequency spectra of the measured microwave when the axial vircator is driven without the ring reflector and with the ring reflector ($D_{A\text{-}RR}$ is 12 mm). The frequency spectra are obtained from the FFT results of the recorded microwave using the high-speed oscilloscope. The overall experimental results with the ring reflector are shown in Table 2.

The microwave power results are normalized based on the microwave power of the axial vircator without the ring reflector to analyze the tendency of microwave enhancement. The normalized microwave power depending on the anode to ring reflector distance is shown in Figure 10. As shown in the figure, the simulations and experiments show analogous tendencies. From the figure, it can be seen that the microwave power of the axial vircator is significantly enhanced when the anode to ring reflector distance is below 21 mm. Both simulations and experiments show that the decay in microwave power enhancement is inversely proportional to the anode to ring reflector distance, and the ring reflector has no enhancing effects after this distance increases to 24 mm. Figure 11 shows the dominant frequency of the simulations and the experiments versus the anode to ring reflector distance. Simulation results show that the dominant frequency of the axial vircator with and without the ring reflector is formed between 5.8 and 6.7 GHz. According to experiments, the dominant frequency is formed between 5.2 and 5.8 GHz. The vircator frequency is proportional to the root of the input voltage and inversely proportional to the A-K gap distance. Because the plateau voltage of the simulations and experiments is the same, the difference in simulation and experimental frequencies is assumed to be caused by variation in the A-K gap distance. Although this distance can be accurately set to 6 mm in simulations, the A-K gap distance can have small variations during the experiments due to installation errors and the surface conditions of the mesh anode and the cathode. Simulations show that the frequency is slightly shifted upward when the anode to ring reflector distance is between 6 and 15 mm. However, according to the experiments, the ring reflector has no distinguishable frequency shifting tendency. From the figures, it is deduced that the power enhancement and the frequency shift are related to the microwave wavelength. The wavelengths corresponding to the average dominant frequency of the simulations and experiments are approximately 50 mm and 54, respectively. The quarter and half wavelengths of the simulations are 12.5 mm and 25 mm, while the quarter and half wavelengths of the experiments are 13.5 mm and 27 mm. If the microwave power enhancement when the anode to ring reflector distance is set to 18 mm is neglected, it is deduced that the ring reflector has a significant influence on the microwave power and the frequency when the anode to ring reflector distance is below the quarter wavelength, and it has no influence when the anode to ring reflector distance is above the half wavelength.

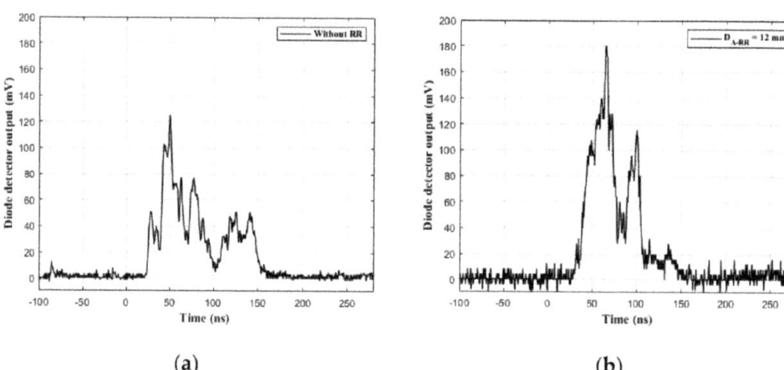

(a) (b)

Figure 8. Typical diode detector output waveform of the axial vircator: (**a**) without ring reflector and (**b**) with ring reflector ($D_{A\text{-}RR}$ = 12 mm).

Table 2. Experimental results.

$D_{A\text{-}RR}$	P_{min} (MW)	P_{max} (MW)	P_{avg} (MW)	Frequency (GHz)
Without RR	8.7	12.3	11.22	5.54
6 mm	18.11	25	22.79	5.79
9 mm	19.36	24.43	21.53	5.59
12 mm	15.14	22.75	20.99	5.48
15 mm	16.22	18.62	18.28	5.64
18 mm	24.38	29.31	25.82	5.48
21 mm	12.53	16.2	15.35	5.16
24 mm	11.22	13.27	12.42	5.57

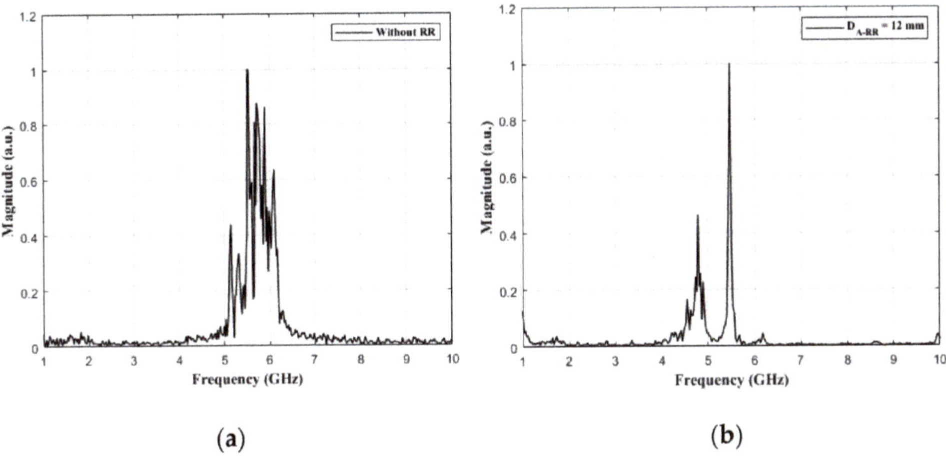

Figure 9. Typical frequency spectrum of the axial vircator: (**a**) without ring reflector and (**b**) with ring reflector ($D_{A\text{-}RR}$ = 12 mm).

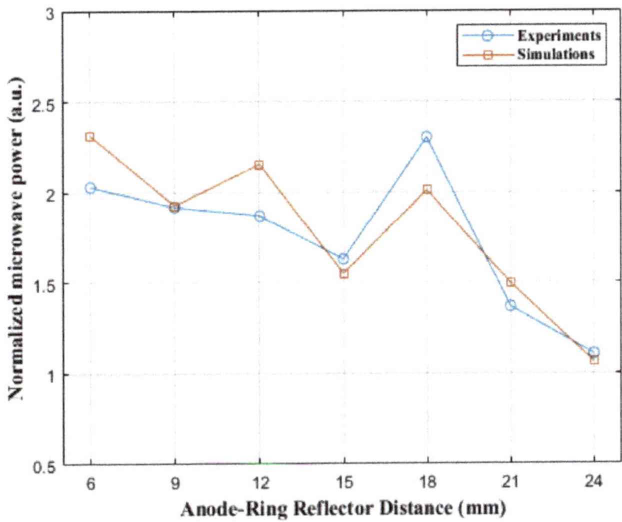

Figure 10. Normalized microwave power of the simulations and experiments as a function of the anode to ring reflector distance.

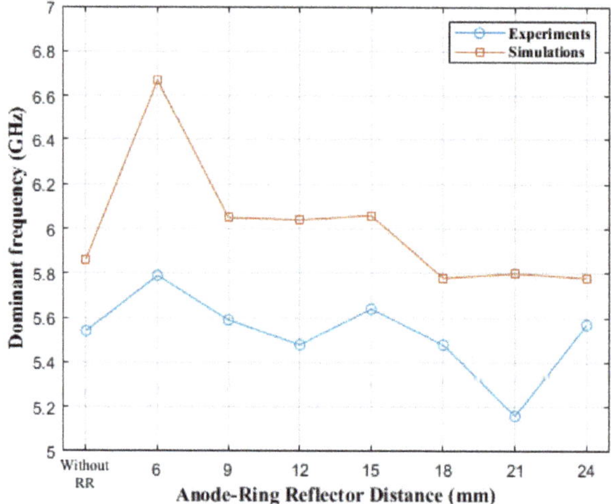

Figure 11. Dominant frequency of the simulations and experiments as a function of the anode to ring reflector distance.

4. Conclusions

In this paper, the power enhancement and the optimum position of the ring reflector for an axial vircator have been investigated using PIC simulations and experiments. The operation features of the axial vircator with the ring reflector are compared with those of the axial vircator without the ring reflector. The axial vircator is driven using a -150 kV voltage pulse from a 10-stage PFN-Marx generator. The ring reflector is installed behind the mesh anode. The anode to ring reflector distance is adjusted in 3 mm steps from 6 mm to 24 mm. When the axial vircator is driven without the ring reflector, the peak microwave power and the dominant frequencies are 11.22 MW and 5.54 GHz, respectively. Installing the ring reflector enhances the microwave power up to 25.82 MW when the anode to ring reflector distance is 18 mm. According to the simulations and experiments, the available working distance between the anode and the ring reflector is limited to below 24 mm. When the anode to ring reflector distance increases above 24 mm, the power enhancement disappears. The difference between the simulations and the experiments is a frequency shift due to the ring reflector position. According to the simulations, the frequency shift is observed when the anode to ring reflector distance is between 6 and 15 mm. Although the frequency shifts upward when the anode to ring reflector distance is 6 mm, the experiments show no noticeable frequency shift when the anode to ring reflector distance is 12 mm or 15 mm. The changes in frequency and the power enhancement due to the position of the ring reflector are assumed to occur at the quarter and half wavelengths of the generated microwave. In future experiments, we intend to investigate the relation between the position of the ring reflector and the wavelength by varying the frequency of the microwave generated by the axial vircator.

Author Contributions: Conceptualization, S.-H.K. and C.-J.L.; methodology, S.-H.K., C.-J.L. and W.-I.K.; validation, S.-H.K., C.-J.L. and W.-I.K.; formal analysis, S.-H.K., C.-J.L. and W.-I.K.; investigation, S.-H.K., C.-J.L. and W.-I.K.; resources, S.-H.K.; data curation, S.-H.K.; writing—original draft preparation, S.-H.K.; writing—review and editing, K.-C.K. All authors have read and agreed to the published version of the manuscript.

Funding: This research received no external funding.

Conflicts of Interest: The authors declare no conflict of interest.

References

1. Benford, J.; Swegle, J.A.; Schamiloglu, E. *High Power Microwaves*, 3rd ed.; CRC Press: Boca Raton, FL, USA, 2015.
2. Jiang, W.; Kristiansen, M. Theory of the virtual cathode oscillator. *Phys. Plasmas* **2001**, *8*, 3781–3787. [CrossRef]
3. Chen, Y.; Mankowski, J.; Walter, J.; Kristiansen, M. Cathode and Anode Optimization in a Virtual Cathode Oscillator. *IEEE Trans. Dielectr. Electr. Insul.* **2007**, *14*, 1037–1044. [CrossRef]
4. Li, L.; Men, T.; Liu, L.; Wen, J. Dynamics of virtual cathode oscillation analyzed by impedance changes in high-power diodes. *J. Appl. Phys.* **2007**, *102*, 123309. [CrossRef]
5. Kovalchuk, B.M.; Polevin, S.D.; Tsygankov, R.V.; Zherlitsyn, A.A. S-Band Coaxial Vircator With Electron Beam Premodulation Based on Compact Linear Transformer Driver. *IEEE Trans. Plasma Sci.* **2010**, *38*, 2819–2824. [CrossRef]
6. Roy, A.; Sharma, A.; Sharma, V.; Patel, A.; Chakravarthy, D.P. Frequency Variation of a Reflex-Triode Virtual Cathode Oscillator. *IEEE Trans. Plasma Sci.* **2013**, *41*, 238–242. [CrossRef]
7. Guo, L.; Shu, T.; Li, Z.; Ju, J.; Fang, X. Preliminary experimental investigation of an X-Band Cerenkov-type high power microwave oscillator without guiding magnetic field. *Rev. Sci. Instrum.* **2017**, *88*, 024708.
8. Neira, E.; Xie, Y.Z.; Vega, F. On the virtual cathode oscillator's energy optimization. *AIP Adv.* **2018**, *8*, 125210. [CrossRef]
9. Mumtaz, S.; Lim, J.S.; Ghimire, B.; Lee, S.W.; Choi, J.J.; Choi, E.H. Enhancing the power of high power microwaves by using zone plate and investigations for the position of virtual cathode inside the drift tube. *Phys. Plasmas* **2018**, *25*, 103113. [CrossRef]
10. Parson, J.M.; Lynn, C.F.; Mankowski, J.J.; Kristiansen, M.; Neuber, A.A.; Dickens, J.C. Conditioning of Carbon Fiber Cathodes in UHV-Sealed Tubes at 200 A/cm^2. *IEEE Trans. Plasma Sci.* **2014**, *42*, 2007–2014. [CrossRef]
11. Parson, J.M.; Lynn, C.F.; Scott, M.C.; Calico, S.E.; Dickens, J.C.; Neuber, A.A.; Mankowski, J.J. A Frequency Stable Vacuum-Sealed Tube High-Power Microwave Vircator Operated at 500 Hz. *IEEE Electron Device Lett.* **2015**, *36*, 508–510. [CrossRef]
12. Andersson, J.; Jansson, M.; Aberg, D. Frequency Dependence of the Anod-Cathode Gap Spacing in a Coaxial Vircator System. *IEEE Trans. Plasma Sci.* **2013**, *41*, 2758–2762. [CrossRef]
13. Roy, A.; Menon, R.; Mitra, S.; Sharma, A.; Nagesh, K.V.; Chakravarthy, D.P. Influence of Electron-Beam Diode Voltage and Current on Coaxial Vircator. *IEEE Trans. Plasma Sci.* **2012**, *40*, 1601–1606. [CrossRef]
14. Turner, G.R. A one-dimentional model illustrating virtual-cathode formation in a novel coaxial virtual-cathode oscillator. *Phys. Plasmas* **2014**, *21*, 093104. [CrossRef]
15. Jeon, W.; Sung, K.Y.; Lim, J.E.; Song, K.B.; Seo, Y.; Choi, E.H. A Diode Design Study of the Virtual Cathode Oscillator with a Ring-Type Reflector. *IEEE Trans. Plasma Sci.* **2005**, *33*, 2011–2016. [CrossRef]
16. Jeon, W.; Lim, J.E.; Moon, M.W.; Jung, K.B.; Park, W.B.; Shin, H.M.; Seo, Y.; Choi, E.H. Output Characteristics of the High-Power Microwave Generated From a Coaxial Vircator With a Bar Reflector in a Drift Region. *IEEE Trans. Plasma Sci.* **2006**, *34*, 937–944. [CrossRef]
17. Gurnevich, E.; Molchanov, P. The Effect of the Electron-Beam Parameter spread on Microwave Generation in a Three-Cavity Axial Vircator. *IEEE Trans. Plasma Sci.* **2015**, *43*, 1014–1017. [CrossRef]
18. Baryshevsky, V.; Gurinovich, A.; Gurnevich, E.; Molchanov, P. Experimental Study of an Axial Vircator with Resonant Cavity. *IEEE Trans. Plasma Sci.* **2015**, *43*, 3507–3511. [CrossRef]
19. Champeaux, S.; Gouard, P.; Cousin, R.; Larour, J. Improved Design of a Multistage Axial Vircator with Reflectors for Enhanced Performances. *IEEE Trans. Plasma Sci.* **2016**, *44*, 31–38. [CrossRef]
20. Baryshevsky, V.; Gurinovich, A.; Gurnevich, E.; Molchanov, P. Experimental Study of a Triode Reflex Geometry Vircator. *IEEE Trans. Plasma Sci.* **2017**, *45*, 631–635. [CrossRef]
21. Barnett, D.H.; Rainwater, K.; Dickens, J.C.; Neuber, A.A.; Mankowski, J.J. A Reflex Triode System with Multicavity Adjustment. *IEEE Trans. Plasma Sci.* **2019**, *47*, 1472–1476. [CrossRef]
22. Zhang, H.; Shu, T.; Liu, S.; Zhang, Z.; Song, L.; Zhan, H. A Compact Modular 5 GW Pulse PFN-Marx Generatro for Driving HPM Source. *Electronics* **2021**, *10*, 545. [CrossRef]
23. Zhang, H.; Yang, J.; Lin, J.; Yang, X. A compact bipolar pulse-forming network-Marx generator based on pulse transformers. *Rev. Sci. Instrum.* **2013**, *84*, 114705. [CrossRef] [PubMed]
24. Verma, R.; Shukla, R.; Sharma, S.K.; Banerjee, P.; Das, R.; Deb, P.; Prabaharan, T.; Das, B.; Mishra, E.; Adhikary, B.; et al. Characterization of High Power Microwave Radiation by an Axially Extracted Vircator. *IEEE Trans. Electron Devices* **2014**, *61*, 141–146. [CrossRef]

 *electronics*

Article

Investigation on a 220 GHz Quasi-Optical Antenna for Wireless Power Transmission

Meng Han [1,2], Xiaotong Guan [2,3], Moshe Einat [4], Wenjie Fu [1,2,*] and Yang Yan [1,2]

1 School of Electronic Science and Engineering, University of Electronic Science and Technology, Chengdu 610054, China; 201711040132@std.uestc.edu.cn (M.H.); yanyang@uestc.edu.cn (Y.Y.)

2 Terahertz Science and Technology Key Laboratory of Sichuan Province, University of Electronic Science and Technology of China, Chengdu 610054, China; guanxt@uestc.edu.cn

3 School of Physics, University of Electronic Science and Technology of China, Chengdu 610054, China

4 Department of Electrical and Electronics Engineering, Ariel University, Ariel 40700, Israel; einatm@ariel.ac.il

* Correspondence: fuwenjie@uestc.edu.cn

Abstract: This paper investigates a 220 GHz quasi-optical antenna for millimeter-wave wireless power transmission. The quasi-optical antenna consists of an offset dual reflector, and fed by a Gaussian beam that is based on the output characteristics of a high-power millimeter-wave radiation source-gyrotron. The design parameter is carried on by a numerical code based on geometric optics and vector diffraction theory. To realize long-distance wireless energy transmission, the divergence angle of the output beam must be reduced. Electromagnetic simulation results show that the divergence angle of the output beam of the 5.6 mm Gaussian feed source has been significantly reduced by the designed quasi-optical antenna. The far-field divergence angle of the quasi-optical antenna in the E plane and H plane is 1.0596° and 1.0639°, respectively. The Gaussian scalar purity in the farthest observation field ($x = 1000$ m) is 99.86%. Thus, the quasi-optical antenna can transmit a Gaussian beam over long-distance and could be used for millimeter-wave wireless power transmission.

Keywords: millimeter waves; wireless power transmitting; quasi-optical antenna; gaussian beam; Gyrotron

Citation: Han, M.; Guan, X.; Einat, M.; Fu, W.; Yan, Y. Investigation on a 220 GHz Quasi-Optical Antenna for Wireless Power Transmission. *Electronics* **2021**, *10*, 634. https://doi.org/10.3390/electronics10050634

Academic Editor: Mikhail Glyavin

Received: 30 January 2021
Accepted: 5 March 2021
Published: 9 March 2021

Publisher's Note: MDPI stays neutral with regard to jurisdictional claims in published maps and institutional affiliations.

1. Introduction

Microwave power transmission (MPT), as a feasible solution for long-distance wireless power transmission, has attracted much attention for many potential applications, such as space solar power generation [1,2], continuous high-altitude relay platforms [3,4], etc. Compared to microwaves, millimeter waves have higher frequencies and better beam directivity [5], which are considered more conducive to long-distance wireless energy transmission applications. However, due to the lack of high efficiency and high power millimeter-wave source (such as magnetron in the microwave region), the investigations on millimeter-wave power transmission are limited.

In recent decades, the high power millimeter-wave source has achieved rapid development [6–8]. Gyrotron, which is also named electron–cyclotron maser, is based on stimulated cyclotron emission processes involving energetic electrons in gyrational motion [9–11]. Unlike the traditional vacuum electronic devices utilizing slow-wave circuits as their interaction structures, gyrotron is a fast-wave device that has much larger physical dimensions than the operating wavelength. Therefore, it has higher output power and higher efficiency than traditional vacuum electronics devices in the millimeter-wave region [12,13]. Until now, a 2.2-MW peak power with an efficiency of 48% has been obtained in 170-GHz gyrotron in Forschungszentrum Karlsruhe, and a 70% peak efficiency with power 0.8-MW has been achieved in 70-GHz gyrotron in the Russian Academy of Science [14]. Thus, the millimeter-wave power transmission based on gyrotron becomes practicable.

However, the divergence angle of the beam output by the gyrotron is large, which makes the energy collapse sharply when it travels a long distance. Therefore, the Gaussian beam output by the gyrotron is mainly applied for short distances [15–17]. Accordingly, to make it suitable for long-distance energy transmission, it is necessary to solve the problem of the large divergence angle of the gyrotron output Gaussian beam. There are still few investigations directly devoted to studying this issue, as far as the author knows, but some similar research work inspire associating. For example, the Gaussian laser beam is different from the Gaussian beam output by the gyrotron in the frequency band. In the application of semiconductor lasers, it has been presented that a collimating lens can be used to increase the waist radius of the Gaussian laser beam, thereby reducing the divergence angle of the Gaussian laser beam [18]. Nevertheless, for a gyrotron with high frequency and high output power, the material and manufacturing process of the required lens are difficult to achieve. In addition, some scholars have proposed a feed-forward Cassegrain geometry to increase the waist radius of the Gaussian laser beam, thereby reducing the divergence angle of the Gaussian laser beam [19]. However, the Gaussian beam obtained by this structure is greatly influenced by the aperture blockage. The center of the gotten beam is empty, and its Gaussian content is not suitable for energy transmission.

Based on the above, this paper proposes an offset Cassegrain dual reflector scheme, which uses 220 GHz gyrotron output Gaussian beam as the feed source to form together with a quasi-optical antenna structure. This design structure can reduce the divergence angle of the Gaussian beam output by the gyrotron, thereby realizing the wireless energy transmission over a long distance in the millimeter-wave band. Meanwhile, 220 GHz is selected as the operating frequency, which is the highest atmospheric window with low transmission attenuation in the millimeter-wave region [5].

The quasi-optical antenna structure and physical principle of reducing the divergence angle of the Gaussian beam output by the gyrotron will be addressed in Section 2, where a numerical calculation program for optimizing the quasi-optical antenna parameters is designed. Simulation results and discussion about the quasi-optical antenna on wireless energy transmission are presented in Section 3, followed by the conclusion in Section 4.

2. Structure and Design Principles

2.1. Antenna Structure

The 3D structure of the proposed quasi-optical antenna is described in Figure 1. And a standard fundamental mode Gaussian feed is considered to be a substitute for the Gaussian beam output by the 220 GHz gyrotron. The dual reflector antenna is composed of a main-reflector and a subreflector. To prevent the feed source and the subreflector from blocking the reflected wave, the offset Cassegrain dual reflector scheme [20,21] is adopted here. The feed source and the subreflector are offset from the main reflector.

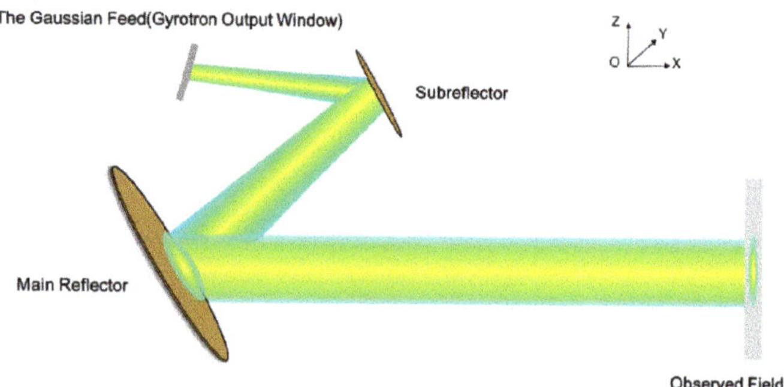

Figure 1. Principle of the quasi-optical antenna and coordinate system.

2.2. Gaussian Beam Propagation Theory

During the propagation of the Gaussian beam, most of the energy is concentrated near the propagation axis. The wave amplitude in the transverse direction is variable and conforms to the Gaussian distribution. Considering the paraxial approximation condition, the field distribution of a fundamental Gaussian beam propagating along the X-direction can be assumed as [22] (pp. 15–16):

$$\varphi(r,x) \;=\; u(r,x)\exp(-jkx) \;=\; \sqrt{\frac{2}{\pi\omega^2(x)}} \times \exp\left(-\frac{r^2}{\omega^2(x)}\right) \times \exp\left(-j\left[k\left(x+\frac{r^2}{2R(x)}\right)\right]-\phi\right) \tag{1}$$

$$\omega(x) \;=\; \omega_0\sqrt{1+\left(\frac{\lambda x}{\pi\omega_0^2}\right)^2} \tag{2}$$

$$R(x) \;=\; x\left[1+\left(\frac{\pi\omega_0^2}{\lambda x}\right)^2\right] \tag{3}$$

where r is the radial distance from the point to the propagation axis X. $\omega(x)$ is defined as the beam radius of the position where the amplitude decreases to $1/e$; At $x = 0$, the beam radius of the Gaussian beam is the smallest, expressed by ω_0, which is also defined as Gaussian beam waist. The wavenumber is $k = 2\pi/\lambda$. $R(x)$ is a measure of the radius of curvature of the wavefront. λ is the wavelength.

It is noteworthy that the Gaussian beam's main mode is an approximate solution under the paraxial condition of the wave equation. Whether this paraxial approximation is effective depends on the electrical size of the Gaussian beam waist. The requirements are as follows [22] (pp. 35–36):

$$\omega_0/\lambda \geq 0.9003 \tag{4}$$

As long as the waist of a Gaussian beam satisfies the above equation, its Gaussian beam propagation characteristics can be guaranteed, and various related formulas can be applied.

Figure 2 shows the propagation of the Gaussian beam and the variation of the beam radius and curvature radius along the propagation direction in the longitudinal section. When the Gaussian beam is far away from the beam waist, the angle between the position where the radius of the Gaussian beam falls to $1/e$ of the maximum value on the x-axis and the z-axis is defined as the far-field divergence angle (half angle) of the Gaussian beam, as follows:

$$\theta \;=\; \tan^{-1}\lim_{x\to\infty}\frac{\omega(x)}{x} \;=\; \tan^{-1}\sqrt{\frac{\lambda}{\pi\omega_0}} \tag{5}$$

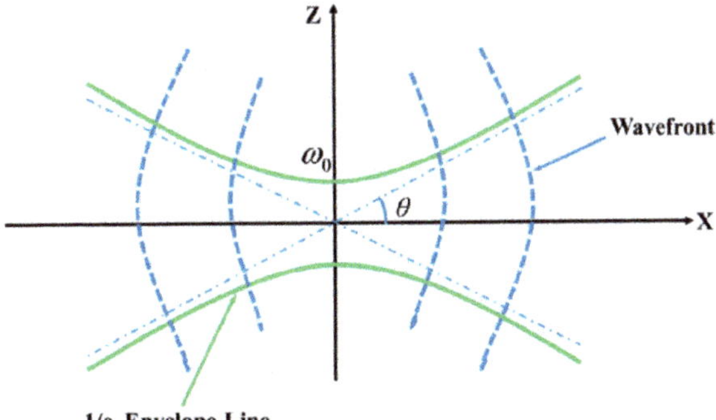

Figure 2. Gaussian beam diagram.

To realize the long-distance bunching transmission of the Gaussian beam, it is necessary to reduce the divergence angle θ by increasing the Gaussian beam waist ω_0. From the perspective of energy, the beam energy contained in the far-field divergence angle accounts for 86.5% [23] of the total energy of the Gaussian beam. Hence, It corresponds to a decrease of 8.68dB in the relative peak power in the far-field pattern.

A Gaussian beam is characterized by the Gaussian mode purity, which is usually described by the correlation coefficient between the output beam E_1 and a theoretical fundamental Gaussian beam E_0 [24]. The correlation coefficient can be defined in two ways: One is the Gaussian scalar content η_s, which involves the amplitude of the field. The other is the Gaussian vector content η_v, including amplitude and phase, which can be expressed as

$$\eta_s = \frac{\iint_S \lfloor E_1 \rfloor \cdot \lfloor E_0 \rfloor dS}{\sqrt{\iint_S \lfloor E_1 \rfloor^2 dS \cdot \iint_S \lfloor E_0 \rfloor^2 dS}} \tag{6}$$

$$\eta_v = \frac{\iint_S E_1^* E_0 dS \cdot \iint_S E_1 E_0^* dS}{\iint_S \lfloor E_1 \rfloor^2 dS \cdot \iint_S \lfloor E_0 \rfloor^2 dS} \tag{7}$$

In millimeter-wave wireless power transmission applications, we mainly consider the Gaussian scalar content of different transmission observation surfaces.

2.3. Design of the Reflectors

The geometry of the classical offset Cassegrain dual reflector antenna [20,21] is shown in Figure 3. To simplify the design, this paper adopts three coordinate systems, including the global coordinate system represented as (O, X, Z), the main-reflector coordinate system depicted as (O_0, X_0, Z_0), and the subreflector coordinate system represented as (O_1, X_1, Z_1).

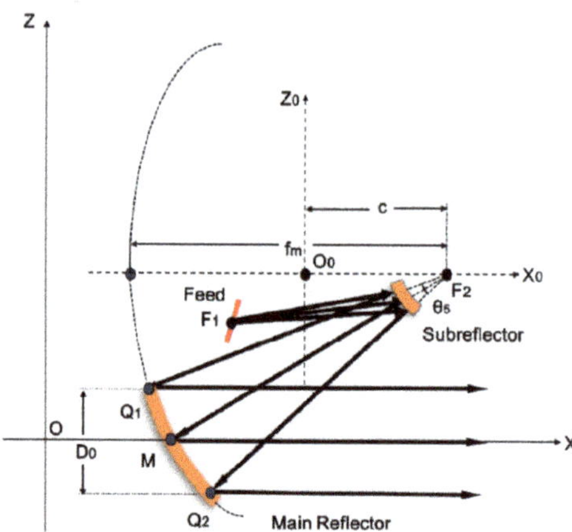

Figure 3. Perspective view of the whole quasi-optical antenna in the X-Z plane.

As shown in Figure 3, the main reflector is a rotating parabolic reflector cut by a cone, which can be expressed in (O_0, X_0, Z_0) coordinate system as

$$x_0 = \frac{z_0^2}{4f_m} - (f_m - c) \quad (z_0 < 0) \tag{8}$$

where f_m is the focal length of the main reflector, and c is the focal length of the subreflector. It is formed by rotating the above part of the parabolic 360° around the focal point F2. The vertex of the cone and the focal point of the parabolic surface are both F2. The cone angle is θ_5.

As illustrated in Figure 4, the subreflector is a rotating hyperboloid surface cut by a cone. The cutting cone of the subreflector is the same as the cutting cone of the main reflector. The cross-section of the second reflecting surface can be expressed in the (O_1, X_1, Z_1) coordinate system as

$$\frac{x_1^2}{a^2} - \frac{(y_1^2 + z_1^2)}{b^2} = 1 \quad (x_1 > 0). \tag{9}$$

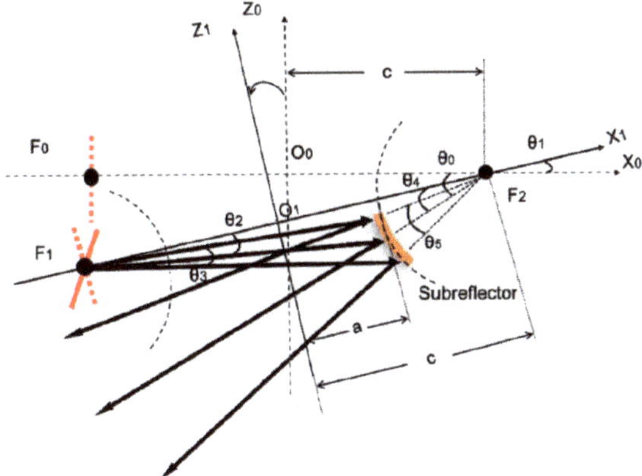

Figure 4. Perspective view of the subreflector in the X_1-Z_1 plane.

It is formed by rotating the right part of the above hyperbola 360° around the focal point F2. The real focus and virtual focus are F1, F2, respectively. The distance from the vertex to the origin O_1 is a and the focal length is c, $b = \sqrt{c^2 - a^2}$.

A fundamental mode Gaussian beam was propagated towards the subreflector and reflected back to the main reflector. According to the geometric optics theory, when the Gaussian ray (emitted from a Gaussian feed source on focus F1) is reflected by the hyperboloid, the reflection line can be regarded as being emitted from the virtual focus F2 of the hyperboloid, which is equivalent to being emitted from the focus F2 of the paraboloid [20]. Therefore, under a certain approximation condition, after being reflected again by the paraboloid, these Gaussian rays can approximately form a beam parallel to the axis of the paraboloid [25]. In other words, the divergent Gaussian beam fed from the focal point F1 can be converted into a plane-like wave—namely, the divergence angle of the Gaussian beam is reduced. Hence, the Gaussian beam can be propagated at a longer distance in free space.

As illustrated in References [20,21], although the dual reflector, shown in Figure 3, can be defined by 21 parameters, only five of them need to be determined for the design, and the other parameters can be derived. However, using only the geometric optics method in [20,21] to obtain the design parameters of the dual reflector is not accurate in the millimeter-wave band. The finite aperture dimension of the dual reflector causes diffraction effects, such as main-reflector spillover, phase error losses, and additional amplitude taper losses [22,26]. Therefore, the vector diffraction theory [27] is applied to accurately verify the performance of the dual reflector.

Based on the vector diffraction theory, the field at any point in free space could be calculated as long as the source field is already known. The observation field radiated from the Gaussian feed can be done by using the Stratton-Chu formula [28].

$$\vec{E}\left(\vec{r}\right) = \oint_{S'} dS' \left\{ i\omega\mu \left[\vec{n} \times \vec{H}\left(\vec{r}'\right)\right] g\left(\vec{r}, \vec{r}'\right) + \left[\vec{n} \cdot \vec{E}\left(\vec{r}'\right)\right] \nabla' g\left(\vec{r}, \vec{r}'\right) + \left[\vec{n} \times \vec{E}\left(\vec{r}'\right)\right] \times \nabla' g\left(\vec{r}, \vec{r}'\right) \right\} \tag{10}$$

$$\vec{H}\left(\vec{r}\right) = \oint_{S'} dS' \left\{ -i\omega\varepsilon \left[\vec{n} \times \vec{E}\left(\vec{r}'\right)\right] g\left(\vec{r}, \vec{r}'\right) + \left[\vec{n} \cdot \vec{H}\left(\vec{r}'\right)\right] \nabla' g\left(\vec{r}, \vec{r}'\right) + \left[\vec{n} \times \vec{H}\left(\vec{r}'\right)\right] \times \nabla' g\left(\vec{r}, \vec{r}'\right) \right\} \tag{11}$$

where $\vec{E}, \vec{H}$ are the electric and magnetic field vectors on the rectangular aperture. μ is the vacuum permeability, ε is the vacuum permittivity, S' is the integral aperture surface, $\vec{n}$ is the unit vector normal to the rectangular aperture surface, and $g\left(\vec{r}, \vec{r}'\right)$ is the point source Green's function, defined by

$$g\left(\vec{r}, \vec{r}'\right) = e^{jk|\vec{r}-\vec{r}'|} / \left(4\pi \left|\vec{r} - \vec{r}'\right|\right) \tag{12}$$

where $\left|\vec{r} - \vec{r}'\right|$ is the distance between the observation point and the source point.

The induction current on the reflector is as follows:

$$\vec{J_E} = 2\left(\vec{n} \times \vec{H}\right). \tag{13}$$

The input power P_{in} in Gaussian feed is normalized, and the received power P_{out} at the observation plane is

$$P_{out} = \iint \frac{1}{2} \mathrm{Re}\left(\vec{E} \times \vec{H}^{\star}\right) \cdot \vec{n} \, dS \tag{14}$$

The calculation of the received power P_{out} on the observation surface is based on the main lobe of the beam obtained on the observation surface. The power transmission efficiency δ_{te} from of the Gaussian feed to the observed plane is

$$\delta_{te} = \frac{P_{out}}{P_{in}} \tag{15}$$

In this paper, a, c, θ_1, θ_3 and f_m are adopted as the initial parameters for the dual reflector, which is feed by a 220 GHz fundamental Gaussian beam with a waist radius of 5.6 mm. The whole quasi-optical antenna is designed and analyzed for wavelength $\lambda = 1.3636$ mm. The initial parameters of the main reflector are $f_m = 140\lambda, c = 55\lambda$. The parameters of the subreflector are $a = 7\lambda, \theta_1 = 20, \theta_3 = 25$. Although other parameters can be derived from formulas in [20,21], considering the diffraction effects, the vector diffraction theory is applied to accurately verify the parameters of the quasi-optical antenna.

Based on the vector diffraction theory, a numerical simulation Matlab code called Gaussian Optical Mirror Transmission (GOMT) was developed to calculate the field distribution on the mirrors and the observed field, which could adjust and optimize the design parameters of the quasi-optical antenna system. The optimization processing is done by adjusting a, c, θ_1, θ_3 and f_m. The field distribution on different output observed planes is calculated. When Gaussian mode purity reaches a satisfying value, for example, the Gaussian scalar content is over 98%, the optimizing work is ended.

Through optimizing the mirror structure parameters by numerical code GOMT, the optimized design parameters are gotten: The final parameter of the main reflector is $f_m = 140\lambda, c = 57.498\lambda$. The final parameter of the subreflector is $a = 7.002\lambda, \theta_1 = 20$, $\theta_3 = 26$. All the final geometrical dimensions, as listed in Table 1, are adjusted by the Matlab code GOMT. The field distribution of the above quasi-optical antenna ($\omega_0 = 5.6$ mm) at 220 GHz calculated by GOMT is demonstrated in Figure 5. It shows that the field radiated from the feed is transformed into a well-shaped Gaussian beam at different output observed fields. Moreover, the overall size of the presented quasi-optical antenna is optimized to be $0.135 \times 0.134 \times 0.210$ mm^3, which is much smaller than the size of the transmitting antenna based on the microwave transmitting system in Reference [29].

Table 1. Geometrical dimensions of the dual reflector.

Parameter	Value	Parameter	Value
Hyperboloid parameter a (mm)	9.5482	θ_3 (°)	26
Hyperboloid parameter c (mm)	78.4063	θ_4 (°)	30
θ_0 (°)	50	θ_5 (°)	32.28
θ_1 (°)	20	f_m (mm)	190.9091
θ_2 (°)	24	D_0 (mm)	132.4000

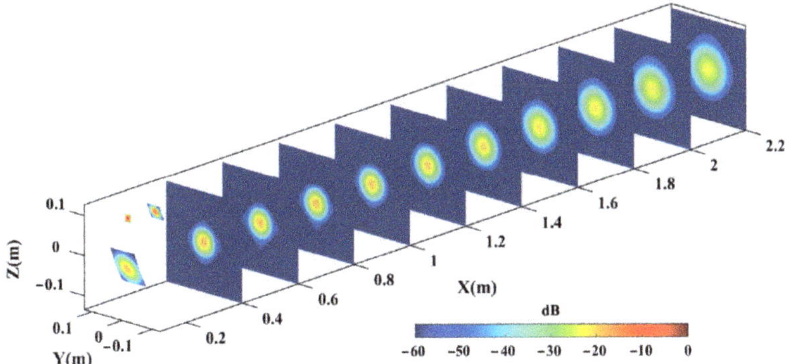

Figure 5. Radiated field distribution of the quasi-optical antenna.

The 3D full-wave simulator Computer Simulation Technology (CST) Microwave Studio has been carried out to verify the GOMT code. The modeling process is as follows: Firstly, setting the center frequency of the whole scheme as f = 220 GHz, we adopt Matlab software to compile a fundamental mode Gaussian beam (ω_0 = 5.6 mm) feed radiation model file; Then, following Equations (8) and (9) and parameters in Table 1, we can determine the dual reflector model structure; Finally, the feed radiation file is imported to excite the dual mirror model, and the Asymptomatic Solver is used to simulate the entire quasi-optical antenna structure.

The results are shown in Figure 6. The obtained waist radius are 39.225 mm and 39.62 mm, respectively. And it is obvious to see that the output electric fields in x = 2.2 m calculated by the CST Microwave Studio commercial software are well consistent with the ones calculated by the GOMT code. For further verification, we compared the Gaussian beam correlation coefficients at different positions of the output observed field. The results are shown in Figure 7. It is clear that the scalar purity calculated by the CST Microwave Studio is in accordance with the one calculated by the GOMT code. This GOMT code was previously applied in designing a quasi-optical mode converter for 220 GHz TE$_{03}$ mode gyrotron, and the experimental results were well consistent with theoretical predictions [30]. The consistency of the CST Microwave Studio commercial software and the GOMT results, as well as the previous experimental verification of the GOMT code, provide solid evidence for the correctness of the GOMT code.

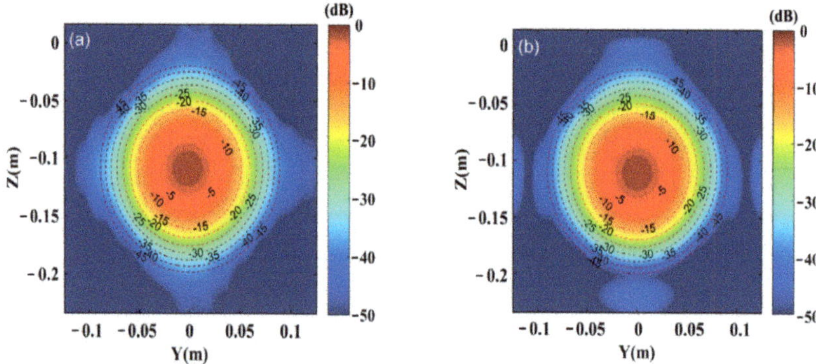

Figure 6. (a) The output observed field ($x = 2.2$ m) calculated by the CST Microwave Studio software; (b) The output observed field ($x = 2.2$ m) calculated by the GOMT code.

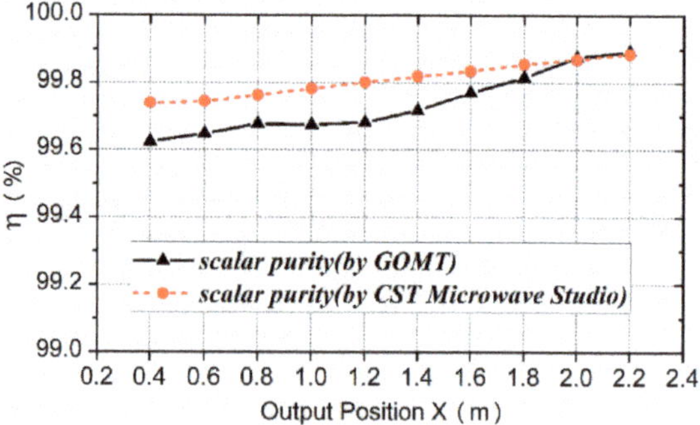

Figure 7. The Gaussian mode content of the output beam was calculated by GOMT and CST Microwave Studio on different output observed fields.

Moreover, compared to the commercial simulation software CST Microwave Studio, the running time is significantly reduced by using the GOMT code. A runtime of seven minutes by GOMT code, compared to more than one day by the 3D full-wave simulation commercial software at the same parameters.

3. Simulation and Discussion

Simulations are initiated by using the 3D full-wave simulator CST Microwave Studio to simulate the reflection and propagation of wave beam through the above designed quasi-optical antenna. The simulation of the cross-section through the quasi-optical antenna in Figure 8b shows the propagation of the energy from the Gaussian feed to the free space on the XZ cross-section. Compared to Figure 8a, the beam radius of the Gaussian beam is significantly reduced; that is, the divergence of the Gaussian beam from the feed source is greatly reduced, which proves the effectiveness of the designed quasi-optical antenna in reducing the Gaussian beam's divergence.

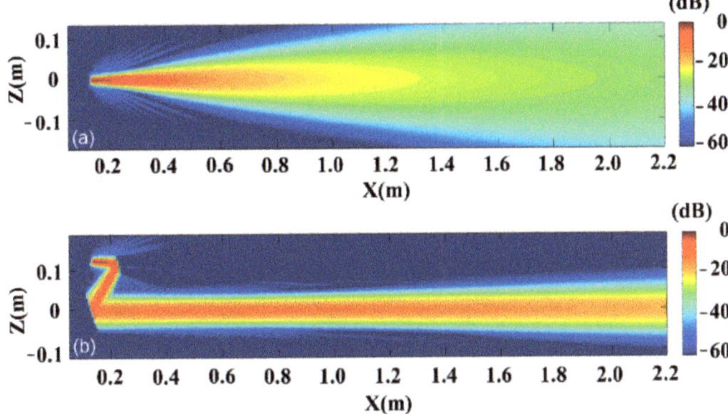

Figure 8. (**a**) Cross-section of the Gaussian-feed with electric field from the 3D simulation; (**b**) Cross-section of the quasi-optical antenna with electric field from the 3D simulation. The contour maps are shown at 20 dB increments from −60 dB to 0.

To further verify the above statements, the far-field radiation patterns of the entire antenna are described in Figure 9. As shown in Table 2, the far-field divergence angle of the 5.6 mm Gaussian feed in E-plane and H-plane is 4.5031° and 4.5032°, respectively. When the dual reflector antenna is added to the 5.6 mm Gaussian feed, the gain of the transmitting system increases from 31.0500 dBi to 43.3442 dBi. However, the far-field divergence angle of E-plane and H-plane decreases by 3.4435° and 3.4393°, respectively. In other words, the far-field divergence angle of the quasi-optical antenna in the E plane and H plane is 1.0596° and 1.0639°, respectively. Therefore, the divergence angle of the output beam of the 5.6 mm Gaussian feed source has been significantly reduced by the designed dual reflector antenna.

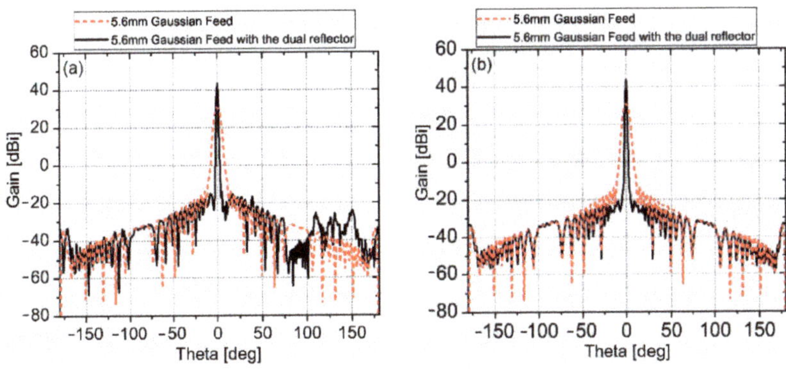

Figure 9. Simulated radiation patterns at 220 GHz for the Gaussian Feed with and without the dual reflector. (**a**) E-plane; (**b**) H-plane.

Table 2. Comparison of the Gaussian feed with and without the Dual Reflector.

Output Observation Position	Gain (dBi)	8.68-dB Beamwidth (°)	
		E-Plane	H-Plane
Gaussian feed with the dual reflector (ω_0 = 5.6 mm)	43.3442	1.0596	1.0639
Gaussian feed (ω_0 = 5.6 mm)	31.0500	4.5031	4.5032
Difference	12.2942	-3.4435	-3.4393

According to Figure 9, the 3-dB beamwidth, sidelobe levels (SLL), and back lobe levels (BLL) of the proposed quasi-optical antenna are summarized in Table 3. Compared with the phased array by the Japanese group [31], the 3-dB beamwidth on the E-plane and H-plane is far behind 7.3°. The gain of the quasi-optical antenna is much higher than 17.6 dBi. Meanwhile, the sidelobe levels and back lobe levels of the proposed quasi-optical antenna on the E-plane and H-plane are very low. This indicates the proposed antenna has a strong anti-interference ability. Moreover, to reduce simulation time, the GOMT code is used to calculate the transmission efficiency at different observation surfaces instead of using a CST Microwave Studio simulator. As shown in Table 4, without considering the atmospheric loss, the power transmission efficiency of different observation surfaces was all above 99%. Besides, as described in Table 5, after using the quasi-optical antenna structure designed above, the waist radius of the Gaussian beam obtained at different observation planes is significantly reduced. Accordingly, the quasi-optical antenna structure is conducive to energy concentration and reception in wireless power transmission.

Table 3. Three-decibel beamwidth, sidelobe level, and backlobe level of the quasi-optical antenna.

	3-dB Beamwidth (°)		SLL (dB)		BLL (dB)			
	E-Plane	H-Plane	E-Plane	H-Plane	E-Plane	H-Plane		
Quasi-Optical Antenna	0.8433	0.8444	-57.2262	-64.3269	-67.84422	-76.5442		

Table 4. Transmission Efficiency of Different Output Observation Positions.

Output Observation Position(m)	0.4	0.6	0.8	1.0	1.2	1.4
Transmission Efficiency δ_{le} (%)	99.574	99.569	99.566	99.562	99.558	99.554
Output Observation Position(m)	1.6	1.8	2.0	2.2	10	1000
Transmission Efficiency δ_{le} (%)	99.546	99.542	99.538	99.534	99.538	99.531

Table 5. Gaussian Beam Waist Radius (GBWR) of the Gaussian feed with and Without the Dual Reflectors at Different Output Observation Positions.

Output Observation Position (m)	0.4	0.6	0.8	1.0	1.2	1.4
GBWR of the Gaussian feed (m)	0.03151	0.04684	0.06226	0.07771	0.09318	0.10866
GBWR of the Gaussian feed with the dual reflector (m)	0.02273	0.02281	0.02331	0.02457	0.02598	0.02832
Output Observation Position (m)	1.6	1.8	2.0	2.2	10	1000
GBWR of the Gaussian feed (m)	0.12414	0.13963	0.15512	0.17062	0.77513	77.5125
GBWR of the Gaussian feed with the dual reflector (m)	0.03037	0.03349	0.03646	0.03923	0.18815	19.4156

Considering for long-distance transmission, the farthest observation surface is set at x = 1000 m, although it is much larger than the far-field condition of the antenna. It can be seen in Figure 10, the Gaussian scalar content transmitted to the observation surface at 1km is up to 99%, and the obtained Gaussian beam waist radius is 19.416 m. The received Gaussian beam waist radius value of 19.416 m is much smaller than the Gaussian beam waist radius value of 77.513 m when only 5.6 mm Gaussian feed is transmitted to x = 1000 m. Thus, the designed quasi-optical antenna has potential in wireless power transmission.

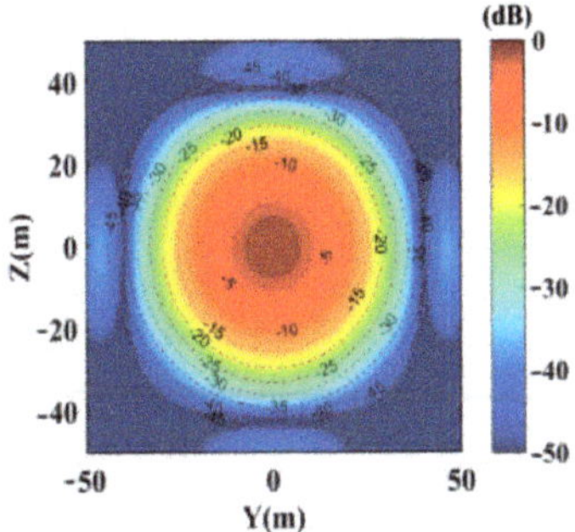

Figure 10. The output observed field in x = 1000 m.

From the above simulation results, it can be concluded that the antenna proposed in this paper has the characteristics of high frequency, small size, high gain, low sidelobe level, small beamwidth, and small divergence angle. These characteristics make it a potential candidate for wireless power transmission in the millimeter-wave frequency band.

As shown in Figure 11a, the far-field divergence angle of the quasi-optical antenna varies with different frequencies. It can be seen that the divergence angle varies at the same level, and the minimum divergence angle is obtained at 220 GHz. Besides, at the furthest observation position, where x = 2.2 m, the simulated quasi-optical antenna Gaussian mode content in the frequency band from 160 to 280 GHz is presented in Figure 11b. The Gaussian scalar mode content of the generated beam is over 99% from 160 to 280 GHz, and the maximum Gaussian scalar mode content with 99.88% is achieved at 180 GHz. Hence, the above results indicate the designed quasi-optical antenna has a very wide operating frequency band in the Gaussian purity and far-field divergence angle.

Next, attempts to explore the relationship between different Gaussian feed waist and the designed quasi-optical antenna transmission system are presented. The specific steps are as follows: When the waist radius ω_0 of the Gaussian beam is constant, according to [32], the electric and magnetic field components of the fundamental Gaussian beam (corresponding to ω_0) can be obtained. Then, the component data are converted into corresponding radiation model files. Considering Equation (4), take the value ω_0 from 2.3 mm to 10 mm, with an interval of 1.1 mm, adopt f = 220 GHz, and its corresponding radiation model files with different Gaussian beam waist can be obtained. Finally, we import different feed radiation files as feed sources to illuminate the dual reflectors designed in this paper.

The simulation results are shown in Figure 12a. On the whole, as the beam waist radius of the input Gaussian feed increases, the Gaussian beam divergence angle becomes larger, but the overall gain shows a downward trend. At frequency f = 220 GHz, the Gaussian feed with the waist (ω_0 = 2.3 mm) and (ω_0 = 3.4 mm) can obtain higher gain and lower divergence angle than other waist radii. At frequency f = 220 GHz, the Gaussian scalar mode content versus the output position and various Gaussian feed waist are shown in Figure 12b, we can see that the Gaussian feed with the waist (ω_0 = 2.3 mm) and (ω_0 = 4.5 mm) can obtain higher Gaussian content than other waist radii at different output observed fields. In other words, considering the antenna gain, far-field divergence angle, and Gaussian scalar content of different observation fields, the Gaussian beam of waist ω_0 = 2.3 mm is the best feed source for the quasi-optical antenna designed in this paper.

Thus, the waist radius of the Gaussian feed has a great impact on the performance of the designed transmitting antenna system, and choosing a suitable Gaussian feed waist radius is important.

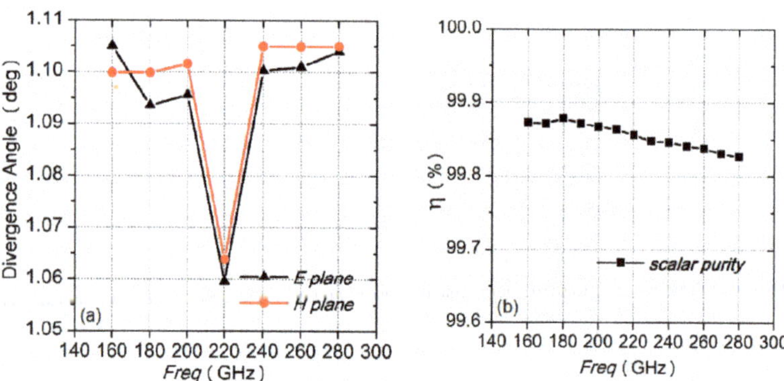

Figure 11. (a) The far-field divergence angle of the quasi-optical antenna versus different frequencies; (b) Gaussian mode purity of the quasi-optical antenna system at the output observed field ($x = 2.2$ m) versus different frequencies.

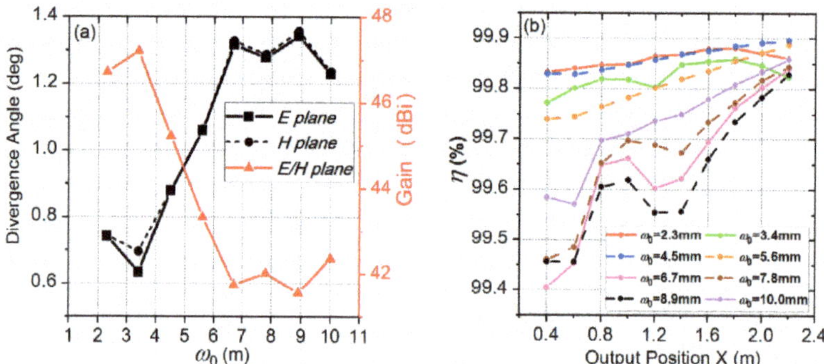

Figure 12. (a) The relation between Gaussian feed waist and divergence angle; the relation between Gaussian feed waist and total gain; (b) Gaussian scalar mode content versus the output position and Gaussian feed waist.

4. Conclusions

The excessive divergence angle of the gyrotron output Gaussian beam is a drawback for its application in wireless power transmission. This paper presented a quasi-optical antenna structure that reduces the divergence angle of the feed and realize long-distance wireless power transmission in the millimeter-wave band. The feed is considered to be a substitute for the Gaussian beam output by the 220 GHz gyrotron.

Considering the diffraction effects, a numerical MATLAB code GOMT is programmed to optimize the design parameters, based on geometric optics and the vector diffraction theory. The numerical code shows similar simulation results with the 3D full-wave simulator CST Microwave Studio, and can sufficiently reduce calculation running time.

The simulation results show the designed quasi-optical antenna has the characteristics of high frequency, small size, high gain, low sidelobe, small beamwidth, small divergence angle, and wide bandwidths. The far-field divergence angles in the E plane and H plane are 1.0596° and 1.0639°, respectively. After using the designed quasi-optical antenna, the waist radius of the Gaussian beam obtained at different observation planes is significantly reduced, facilitating the transmission and collection of energy. Even transmitted to 1 km distance, the radiation kept Gaussian distribution well and the Gaussian scalar content is up to 99%, and the obtained Gaussian beam waist radius is 19.416 m.

Additionally, the simulation indicated that different input Gaussian feed waists have a great impact on the performance of the quasi-optical antenna. Considering the factors of antenna gain, far-field divergence angle, and Gaussian scalar content of different observation fields, the Gaussian beam of the waist $\omega_0 = 2.3$ mm is the best feed source for the quasi-optical antenna designed in this paper. Hence, the research results also could provide references and requirements for developing gyrotron for better long-distance millimeter-wave power transmission. A future step will optimize the output beam quality of the gyrotron to satisfy this requirement.

Author Contributions: For research articles with several authors, M.H. and W.F. contributed to the overall study design, analysis, computer simulation, and writing of the manuscript. M.E. and W.F. contributed to the conception, X.G., M.E. and Y.Y. provided technical support and revised the manuscript. All authors have read and agreed to the published version of the manuscript.

Funding: This work was supported by the National Key Research and Development Program of China under 2019YFA0210202, the National Natural Science Foundation of China under Grant 61971097, 6201101342, the Sichuan Science and Technology Program under Grant 2018HH0136, and the Terahertz Science and Technology Key Laboratory of Sichuan Province Foundation under Grant THZSC201801.

Institutional Review Board Statement: Not applicable.

Informed Consent Statement: Not applicable.

Data Availability Statement: Data sharing not applicable.

Acknowledgments: In this section you can acknowledge any support given which is not covered by the author contribution or funding sections. This may include administrative and technical support, or donations in kind (e.g., materials used for experiments).

Conflicts of Interest: The authors declare no conflict of interest.

References

1. Glaser, P.E. Power from the Sun: Its Future. *Science* **1968**, *162*, 857–861. [CrossRef]
2. Shinohara, N.; Kawasaki, S. Recent Wireless Power Transmission technologies in Japan for space solar power station/satellite. In Proceedings of the 2009 IEEE Radio and Wireless Symposium, San Diego, CA, USA, 18–22 June 2009; pp. 13–15.
3. Schlesak, J.; Alden, A.; Ohno, T. SHARP rectenna and low altitude flight trials. In Proceedings of the GLOBECOM'85-Global Telecommunications Conference, New York, NY, USA, 2–5 December 1985; pp. 960–964.
4. East, T.W. A self-steering array for the SHARP microwave-powered aircraft. *IEEE Trans. Antennas Propag.* **1992**, *40*, 1565–1567. [CrossRef]
5. Button, J.K.; Wiltse, J.C. *Infrared and Millimeter Waves V4: Millimeter Systems*; Elsevier: Amsterdam, The Netherlands, 2014; pp. 1–2.
6. Gilmour, A.S. *Microwave and Millimeter-Wave Vacuum Electron Devices: Inductive Output Tubes, Klystrons, Traveling-Wave Tubes, Magnetrons, Crossed-Field Amplifiers, and Gyrotrons*; Artech House: Boston, MA, USA, 2020.
7. Barker, R.J.; Luhmann, N.C.; Booske, J.H.; Nusinovich, G.S. *Modern Microwave and Millimeter-Wave Power Electronics*; Institute of Electrical and Electronics Engineers (IEEE): Hoboken, NJ, USA, 2005.
8. Chatterjee, R. Microwave and Millimeter-wave Vacuum Tube Electron Devices: Overview and State-of-the-art. *IETE Tech. Rev.* **1993**, *10*, 175–181. [CrossRef]
9. Chu, K.R. The electron cycltron maser. *Rev. Modern Phys.* **2004**, *76*, 489–540. [CrossRef]
10. Schneider, J. Stimulated Emission of Radiation by Relativistic Electrons in a Magnetic Field. *Phys. Rev. Lett.* **1959**, *2*, 504–505. [CrossRef]
11. Hirshfield, J.L.; Wachtel, J.M. Electron Cyclotron Maser. *Phys. Rev. Lett.* **1964**, *12*, 533–536. [CrossRef]
12. Flyagin, V.A.; Gaponov, A.V.; Petelin, I.; Yulpatov, V.K. The Gyrotron. *IEEE Trans. Microw. Theory Tech.* **1977**, *25*, 514–521. [CrossRef]
13. Nusinovich, G.S.; Thumm, M.K.A.; Petelin, M.I. The Gyrotron at 50: Historical Overview. *J. Infrared Millim. Terahertz Waves* **2014**, *35*, 325–381. [CrossRef]

14. Thumm, M. State-of-the-Art of High-Power Gyro-Devices and Free Electron Masers. *J. Infrared Millim. Terahertz Waves* **2020**, *41*, 1–140. [CrossRef]
15. Prinz, O.; Arnold, A.; Gantenbein, G.; Liu, Y.-H.; Thumm, M.; Wagner, D. Highly Efficient Quasi-Optical Mode Converter for a Multifrequency High-Power Gyrotron. *IEEE Trans. Electron Devices* **2009**, *56*, 828–834. [CrossRef]
16. Thumm, M.; Yang, X.; Arnold, A.; Dammertz, G.; Michel, G.; Pretterebner, J.; Wagner, D. A High-Efficiency Quasi-Optical Mode Converter for a 140-GHz 1-MW CW Gyrotron. *IEEE Trans. Electron Devices* **2005**, *52*, 818–824. [CrossRef]
17. Zhao, G.; Xue, Q.; Wang, Y.; Wang, X.; Zhang, S.; Liu, G.; Feng, J.; Zhang, L. Design of Quasi-Optical Mode Converter for 170-GHz TE32,9-Mode High-Power Gyrotron. *IEEE Trans. Plasma Sci.* **2019**, *47*, 2582–2589. [CrossRef]
18. Zhou, X.-Q.; Ann, B.N.K.; Seong, K.S. Single aspherical lens for deastigmatism, collimation, and circularization of a laser beam. *Appl. Opt.* **2000**, *39*, 1148–1151. [CrossRef] [PubMed]
19. Shen, C.Y.; Chen, F.; Yu, X.D. Optical design of an optical system for free space optical communication. In Proceedings of the Advanced Optical Manufacturing Technologies—International Symposium on Advanced Optical Manufacturing & Testing Technologies, Xian, China, 23 February 2006; p. 614915.
20. Rusch, W.; Prata, A.; Rahmat-Samii, Y.; Shore, R. Derivation and application of the equivalent paraboloid for classical offset Cassegrain and Gregorian antennas. *IEEE Trans. Antennas Propag.* **1990**, *38*, 1141–1149. [CrossRef]
21. Granet, C. Designing axially symmetric Cassegrain or Gregorian dual-reflector antennas from combinations of prescribed geometric parameters. *IEEE Antennas Propag. Mag.* **1998**, *40*, 76–82. [CrossRef]
22. Goldsmith, P.F. *Quasioptical Systems: Gaussian Beam Quasioptical Propagation and Applications*; IEEE Press: Piscataway, NJ, USA, 1998.
23. Wang, L. *Research on Quasi-Optical Technology in Millimeter Wave Space Beam Power Combining [Dissertation]*; Southeast University: Nanjing, China, 2018.
24. Jin, J.; Piosczyk, B.; Thumm, M.; Rzesnicki, T.; Zhang, S. Quasi-Optical Mode Converter/Mirror System for a High-Power Coaxial-Cavity Gyrotron. *IEEE Trans. Plasma Sci.* **2006**, *34*, 1508–1515. [CrossRef]
25. Xie, Z.M.; Wang, L.X. A linear horn array feed dual reflector antenna for spatial power combining. In Proceedings of the 2012 International Conference on Microwave and Millimeter Wave Technology (ICMMT), Shengzhen, China, 5–8 May 2012; pp. 1–4.
26. Kogelnik, H.; Li, T. Laser beams and resonator. *Appl. Opt.* **1966**, *5*, 1312–1325. [CrossRef]
27. Marathay, A.S.; McCalmont, J.F. Vector diffraction theory for electromagnetic waves. *J. Opt. Soc. Am. A* **2001**, *18*, 2585–2593. [CrossRef] [PubMed]
28. Kong, J.A. *Electromagnetic Wave Theory*; Wiley: New York, NY, USA, 1986; p. 381.
29. Chen, C.; Huang, K.; Yang, Y. Microwave Transmitting System Based on Four-Way Master-Slave Injection-Locked Magnetrons and Horn Arrays With Suppressed Sidelobes. *IEEE Trans. Microw. Theory Tech.* **2018**, *66*, 2416–2424. [CrossRef]
30. Zhang, C.X.; Fu, W.J.; Yan, Y. Study on a gyrotron quasi-optical mode converter for terahertz imaging. *J. Electromagn. Waves Appl.* **2021**, *35*, 176–184. [CrossRef]
31. Matsumoto, H. Research on solar power satellites and microwave power transmission in Japan. *IEEE Microw. Mag.* **2002**, *3*, 36–45. [CrossRef]
32. Haus, H.A. *Waves and Fields in Optoelectronics*; Prentice-Hall: Englewood Cliffs, NJ, USA, 1984; pp. 113–115.

electronics

Article

Frequency Tuning Characteristics of a High-Power Sub-THz Gyrotron with Quasi-Optical Cavity

Xiaotong Guan [1,2], Jiayi Zhang [2,3], Wenjie Fu [2,3,*], Dun Lu [2,3], Tongbin Yang [2,3], Yang Yan [2,3] and Xuesong Yuan [2,3]

[1] School of Physics, University of Electronic Science and Technology of China, Chengdu 610054, China; guanxt@uestc.edu.cn

[2] Terahertz Science and Technology Key Laboratory of Sichuan Province, University of Electronic Science and Technology of China, Chengdu 610054, China; zjy@std.uestc.edu.cn (J.Z.); ludun@std.uestc.edu.cn (D.L.); yangtongbin@std.uestc.edu.cn (T.Y.); yanyang@uestc.edu.cn (Y.Y.); yuanxs@uestc.edu.cn (X.Y.)

[3] School of Electronic Science and Engineering, University of Electronic Science and Technology of China, Chengdu 610054, China

* Correspondence: fuwenjie@uestc.edu.cn

Abstract: Motivated by some emerging high-frequency applications, a high-power frequency-tunable sub-THz quasi-optical gyrotron cavity based on a confocal waveguide is designed in this paper. The frequency tuning characteristics of different approaches, including magnetic field tuning, mirror separation adjustment, and hybrid tuning, have been investigated by particle-in-cell (PIC) simulation. Results predict that it is possible to realize a smooth continuous frequency tuning band with an extraordinarily broad bandwidth of 41.55 GHz, corresponding to a relative bandwidth of 18.7% to the center frequency of 0.22 THz. The frequency tunability is provided by varying the separation distance between two mirrors and correspondingly adjusting the external magnetic field. During the frequency tuning, the output power remains higher than 20 kW, which corresponds to an interaction efficiency of 10%. Providing great advantages in terms of broad bandwidth, smooth tuning, and high power, this research may be conducive to the development of high-power frequency-tunable THz gyrotron oscillators.

Keywords: gyrotron; quasi-optical cavity; confocal waveguide; frequency tuning; high power; sub-millimeter wave; terahertz

Citation: Guan, X.; Zhang, J.; Fu, W.; Lu, D.; Yang, T.; Yan, Y.; Yuan, X. Frequency Tuning Characteristics of a High-Power Sub-THz Gyrotron with Quasi-Optical Cavity. *Electronics* **2021**, *10*, 526. https://doi.org/10.3390/electronics10050526

Academic Editor: Mikhail Glyavin

Received: 30 January 2021
Accepted: 19 February 2021
Published: 24 February 2021

Publisher's Note: MDPI stays neutral with regard to jurisdictional claims in published maps and institutional affiliations.

1. Introduction

A gyrotron is a typical fast-wave vacuum electron device based on the interaction principle between gyrating electrons and the electromagnetic waves propagating in the waveguide [1]. As one of the most powerful radiation sources, a gyrotron performs with the capability of high-power output from the microwave to terahertz (THz) band [2,3]. Up till now, the world power record for a gyrotron is 2.2 MW at 170 GHz applied for electron cyclotron heating and current drive in the International Thermonuclear Experimental Reactor (ITER) [4]. In recent decades, a continuous frequency-tunable gyrotron operating at a single mode has been especially attractive for some modern high-frequency applications [5], such as high-resolution molecular gas spectroscopy [6], nuclear magnetic resonance spectroscopy enhanced by dynamic nuclear polarization (DNP-NMR) [7], and the direct measurement of positronium hyperfine splitting (Ps-HFS) [8], in which radiation sources are required to be high-power and continuously tunable in a wide frequency range.

According to the principles of an electron cyclotron maser (ECM), the gyrating electrons are able to interact with the electromagnetic (EM) waves efficiently only under the cyclotron resonance condition [2].

$$\omega - k_z v_z \approx s\Omega_c, \quad \Omega_c = \frac{eB_0}{\gamma m_0}, \quad \gamma = 1 + \frac{eV_0}{m_0 c^2} \tag{1}$$

where ω and k_z are the angular frequency and axial wavenumber of the EM waves in the interaction space (cavity); v_z is the electron axial velocity; Ω_c is the relativistic cyclotron frequency of the electrons relative to the static magnetic field strength B_0, the accelerate voltage V_0, and the cyclotron harmonic number s; γ is the relativistic factor of the electrons; m_0 and e are the relativistic electron rest mass and charge; and c is the velocity of light in free space.

From Equation (1), there are two possible approaches to controlling the output frequency: changing the cyclotron frequency Ω_c of the electrons, or changing the EM waves' frequency ω in the cavity. The variation of Ω_c can be easily achieved by altering the magnetic field B_0 or the beam voltage V_0. However, to meet the requirement of high-power output, a conventional gyrotron normally employs a high Q-value cavity (in the order of ~1000), which significantly restricts its resonance bandwidth ($\Delta f \sim f_0/Q$). That is why conventional high-power gyrotrons can only achieve a narrowband frequency by tuning at a fixed mode, or discrete broadband tuning for several modes. Therefore, the key issue with frequency-tunable gyrotrons is broadening the cavity resonance bandwidth as much as possible without reducing its Q-value.

For a conventional gyrotron with a cylindrical cavity, several approaches have been proposed and demonstrated for the problem of frequency tuning. One of the most important mechanisms relies on exciting a series of high-order axial modes (HOAMs) in a long gyrotron cavity. With a well-elaborated selection of cavity length and beam current, the frequency region obtained by operating in one axial mode is able to overlap with the frequency region in another axial mode [9]. In principle, continuous broadband frequency tuning can be accessible for gyrotron operation by increasing the axial mode indices. Nowadays, lots of continuous frequency-tunable THz gyrotrons have been successfully developed and applied to DNP-NMR applications at the Massachusetts Institute of Technology (MIT, Cambridge, MA, USA) [10], the Bruker Biospin company in collaboration with the Communications & Power Industries Company (CPI, Palo Alto, CA, USA) [11], and the Research Center for the Development of the Far-Infrared Region of University of Fukui (FIR-UF, Fukui, Japan) [12]. However, to maintain a wide bandwidth, the operating beam currents for these HOAMs are limited at several hundred milliamperes, resulting in a medium power level (less than 100 W). There is the same weakness in other frequency tuning methods, including using an improved multi-section cavity [13], cathode-end power output [14], and using backward-wave components [15]. Recently, the Terahertz Research Center of University of Electronic Science and Technology of China (TRC-UESTC) reported a frequency-tunable HOAMs gyrotron operating at a high beam current and with experimental output higher than 0.45 kW over a 0.79 GHz frequency range [16], which was high-power but not continuously tunable.

As for the gyrotron cavity based on a multi-conductor waveguide, there is another approach for acquiring smooth frequency tunability by adjusting its structural parameters. For example, in a coaxial gyrotron cavity, the cavity eigenfrequency depends on the ratio of the radii of the external and internal conductors. It is possible to realize a continuous frequency tuning by moving the tapered inner conductor longitudinally. Researchers at the Institute of Applied Physics of the Russian Academy of Sciences (IAP, Moscow, Russia) have numerically investigated this tuning mechanism. One result presented the possibility of frequency tuning by 8 GHz at around 394.6 GHz (within a frequency band of about 2%), with an output power of about several hundred watts [17], while another result demonstrated the smooth frequency tuning at one mode by no less than 3.5% around 330 GHz with about 10 kW output power [18]. Lately, a similar mechanical tuning method was applied to a high-power large orbit gyrotron (LOG) with a variable slit-cavity [19]. By changing the slit width transversely, theoretical calculation results suggested a relative frequency bandwidth of 8.5% around 316 GHz with an output power of over 10 kW.

Recently, high-power continuous frequency-tunable radiation over a 3.1 GHz bandwidth around 0.2 THz has been experimentally observed in a quasi-optical gyrotron with a straight confocal waveguide [20], which does not contain a resonance structure. Although lacking a comprehensive theoretical model to explain it, the experimental result still points out the possibility of generating high-power frequency-tunable THz radiation from a quasi-optical gyrotron with a confocal waveguide.

A quasi-optical cavity provides many good characteristics, such as a high power capacity and a low mode density, which bring positive effects to high-frequency and high-power gyrotron design. In the 1990s, a quasi-optical gyrotron based on a Fabry–Perot cavity generated an output power of 90 kW at 100 GHz, operated at the fundamental cyclotron [21]. The cylindrical confocal waveguide is another quasi-optical structure. Utilizing a confocal waveguide as the interaction structure for a gyrotron was first proposed at MIT, and experimentally demonstrated later by a 140 GHz fundamental gyrotron oscillator [22] and a 140 GHz gyrotron traveling wave amplifier (gyro-TWA) [23]. As for the harmonic gyrotron, a 0.4 THz gyrotron with a confocal cavity was developed at TRC-UESTC and, experimentally, achieved an output power of 6.44 kW operating at the second cyclotron harmonic [24].

Furthermore, as a type of open structure, a quasi-optical waveguide presents a natural frequency tunability related to the separation distance between two mirrors, which introduces another method for frequency tuning. In this paper, we propose a high-power broadband continuous frequency-tunable gyrotron cavity based on a confocal waveguide. Its frequency tuning characteristics fall under three different strategies, namely magnetic field tuning, mirror separation adjustment, and the hybrid tuning of the above two parameters, which are both investigated by particle-in-cell (PIC) simulation. Results suggest that the proposed quasi-optical cavity is able to generate high power of no less than 20 kW over a smoothly continuous frequency tuning band with an extraordinarily broad bandwidth of 41.55 GHz around 0.22 THz. Compared with other frequency tuning approaches, this method provides advantages in terms of high power, broad band, and smooth continuity.

The paper is organized as follows: the design principles of the quasi-optical cavity, including the introduction of a quasi-optical waveguide, the cavity design, and the cold cavity frequency tunability, are described in Section 2. Section 3 presents the detailed PIC simulation results for three different frequency tuning approaches, covering magnetic tuning, mirror separation adjustment, and hybrid tuning. Finally, a summary for this work is reported in Section 4.

2. Cavity Design Principles

2.1. Quasi-Optical Waveguide

As shown in Figure 1, the open quasi-optical waveguide is composed of two identical cylindrical mirrors with a finite aperture of $2a$ and a curvature radius of R_c. When the separation distance between the two mirrors $L_\perp$ is equal to R_c, the two mirrors form a confocal system, which is called a confocal waveguide.

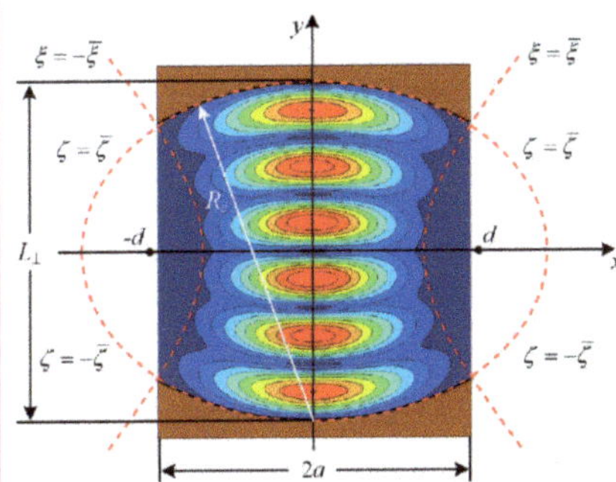

Figure 1. Cross-section scheme of the quasi-optical waveguide and the electric field distribution for the TE_{06} mode.

Under the geometrical optical approximation for a high frequency wave, the membrane function for the transverse electric (TE) mode in the open waveguide can be obtained by solving the Helmholtz equation in the elliptic coordinate system [24]. The numerically calculated results of the electron field distribution for the TE_{06} mode are shown in Figure 1. The eigen-wavenumber k_{mn} and the cut-off frequency f_{cmn} for the TE_{mn} mode can be deduced as

$$k_{mn} = \frac{\pi}{L_\perp}\left(n + \frac{2m+1}{\pi}\arcsin\sqrt{\frac{L_\perp}{2R_c}}\right) \tag{2}$$

$$f_{cmn} = \frac{k_{mn}c}{2\pi} = \frac{c}{2L_\perp}\left(n + \frac{2m+1}{\pi}\arcsin\sqrt{\frac{L_\perp}{2R_c}}\right) \tag{3}$$

According to Equations (2) and (3), the eigenfrequency for the EM mode in a quasi-optical waveguide is almost linear with the mirror separation distance $L_\perp$. It appears that a smoothly continuous variation of the operating frequency can be easily realized in a quasi-optical gyrotron cavity. By moving the mirrors smoothly, the cut-off frequency for the operating mode will change, leading to the possibility of the continuous tuning of the oscillation frequency in the quasi-optical cavity. This brings a new approach to frequency tunability that cannot be accomplished in traditional gyrotrons based on closed waveguides.

On the other hand, the quasi-optical waveguide provides an impressive mode selection feature. Since lacking sidewalls, as shown in Figure 1, some EM modes will be diffracted out and undergo a large diffraction loss. Previous researchers have demonstrated that only the TE_{0n} mode could be stably propagated by selecting a small mirror aperture. Thus, the frequency separation between neighboring modes in a quasi-optical waveguide is about $\Delta f = c/(2L_\perp)$, which is a much greater isolation than that of a cylindrical or coaxial waveguide. The lower mode density provides a distinct advantage for quasi-optical gyrotrons in realizing broadband frequency tuning.

2.2. Frequency Tuning Characteristics in a Cold Cavity

As shown in Figure 2, a TE_{06} mode frequency-tunable quasi-optical cavity for a high-power sub-THz gyrotron is designed and studied in this paper. This cavity is similar to the 0.4 THz second harmonic confocal cavity reported previously [24]. The mirror radius in the straight section is set to 4.20 mm, corresponding to a cut-off frequency of 223.06 GHz for a TE_{06} mode under a rigorously confocal situation. The detailed structural parameters of the designed quasi-optical cavity are listed in Table 1.

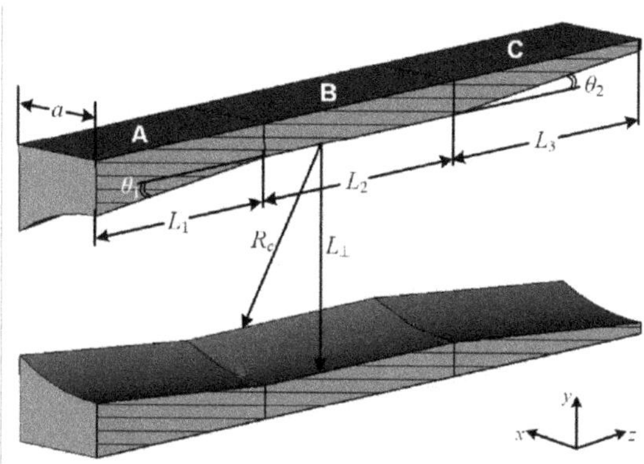

Figure 2. Structural configuration of the quasi-optical gyrotron cavity.

Table 1. Structural parameters of the proposed quasi-optical gyrotron cavity.

R_c	$L_\perp$	a	L_1	L_2	L_3	θ_1	θ_2
4.20 mm	4.20 mm [1]	2.2 mm	10 mm	13 mm	10 mm	2.12°	2.29°

[1] Under a rigorously confocal condition.

For a cold cavity with the absence of an electron beam, the oscillation frequency f_{osc} can be obtained by solving the following differential equation

$$\frac{d^2 f(z)}{dz^2} + k_z^2(z) f(z) = 0 \tag{4}$$

satisfying certain boundary conditions at the left end ($z = z_{in}$) and the right end ($z = z_{out}$).

$$\left. \frac{df(z)}{dz} \right|_{z=z_{in}} - jk_z(z_{in}) f(z_{in}) = 0 \tag{5}$$

$$\left. \frac{df(z)}{dz} \right|_{z=z_{out}} + jk_z(z_{out}) f(z_{out}) = 0 \tag{6}$$

Based on the complex oscillation frequency f_{osc} and the axial field distribution $f(z)$, the quality factor Q for the metallic quasi-optical cavity can be calculated by

$$\frac{1}{Q} = \frac{1}{Q_{diff\parallel}} + \frac{1}{Q_{diff\perp}} + \frac{1}{Q_{ohm}} \tag{7}$$

$$Q_{diff\parallel} = \frac{\text{Re}(f_{osc})}{2\text{Im}(f_{osc})}, \ Q_{diff\perp} = \frac{k_{mn}L_\perp}{\Lambda}, \ Q_{ohm} = \frac{2}{\delta} \frac{\iiint_V |H|^2 dV}{\iint_{S_{mirror}} |H_\perp|^2 dS} \tag{8}$$

where δ is the skin depth on the metallic mirror surface S_{mirror}, and the diffraction loss parameter Λ can be approximated as

$$\log_{10} \Lambda = -0.0069 C_F^2 - 0.7088 C_F + 0.5443, \ \text{for TE}_{0n} \text{ mode}, \ C_F = k_{mn}a^2/L_\perp \tag{9}$$

For a rigorously confocal cavity where $L_\perp = R_c = 4.20$ mm, the cold cavity characteristics for axial modes TE$_{06q}$ (q = 1, 2, 3, 4) are numerically calculated. The cold cavity oscillation frequency f, the Q-value, and the frequency band for each TE$_{06q}$ mode are listed in Table 2, while Figure 3 shows their normalized axial field profiles. Theoretical results suggest that the output frequency of the confocal cavity may be tuned in a frequency range of 4.33 GHz, from 223.20 to 227.53 GHz, if the first four HOAMs can be excited. It is a wider band compared to the previously reported HOAMs gyrotron with a cylindrical cavity, in Reference [16].

Table 2. Cold cavity characteristics for the TE$_{06q}$ modes.

q	Frequency f (GHz)	Q	$f - f/Q$ (GHz)	$f + f/Q$ (GHz)
1	223.27	3087	223.198	223.342
2	223.91	779	223.623	224.197
3	224.98	350	224.337	225.623
4	226.44	208	225.351	227.529

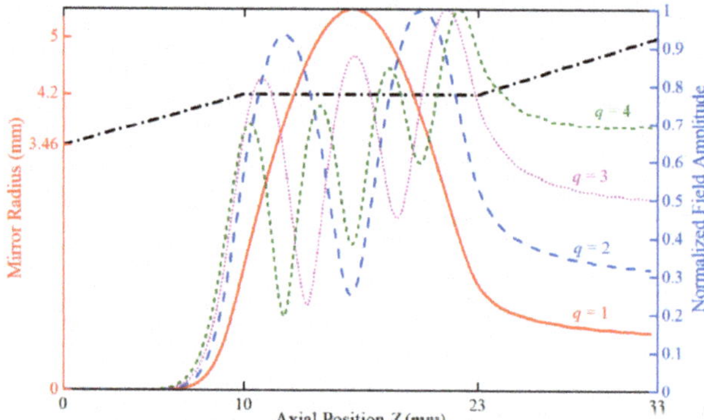

Figure 3. Calculated axial field profiles of the TE_{06q} ($q = 1, 2, 3, 4$) modes in the cold cavity.

Besides, the dependencies of the cold cavity oscillation frequency and the Q-value for the TE_{061} mode on the mirror separation distance $L_\perp$ are calculated and plotted in Figure 4. It should be pointed out that the adjusting of mirror separation $L_\perp$ in this work is improved by moving the total upper and lower parts of the cavity with geometrical symmetry in the y-direction, rather than just the straight sections reported in References [25,26].

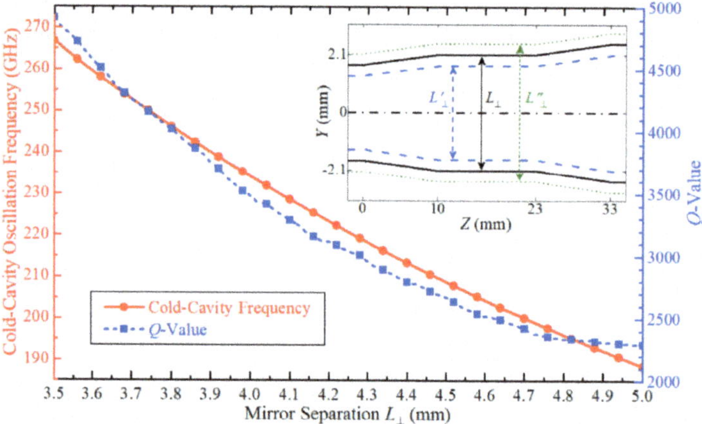

Figure 4. Calculated results of the cold cavity oscillation frequency and the Q-value for the operating TE_{061} mode depending on the mirror separation distance $L_\perp$. The adjustment method of $L_\perp$ is schematically shown in the illustration.

As illustrated in Figure 4, the calculation results predict that the cold cavity oscillation frequency can be smoothly varied by adjusting the separation distance $L_\perp$. If $L_\perp$ changes from 3.80 to 4.60 mm, within a ± 0.40 mm displacement compared to the confocal condition $L_\perp = 4.20$ mm, the oscillation frequency for the TE_{061} mode will be tuned from 246.14 to 204.38 GHz continuously, corresponding to a significant wideband of 41.76 GHz (about 19% around 0.22 THz). At the same time, the cavity's Q-value will gradually reduce from 4016 to 2520, but still be retained at a high level. The great frequency characteristics of the cold cavity make it possible to achieve broadband frequency tuning in gyrotron operations.

3. PIC Simulation

To investigate the output performance of the proposed quasi-optical cavity driven by a gyrating electron beam, especially for the frequency tuning characteristics, the designed model has been built and simulated with the help of a 3D particle-in-cell (PIC) code, CHIPIC [27]. The initial operating parameters of the electron beam are listed in Table 3, and are based on the linear gyrotron theory [24]. During the simulation procedure, the beam velocity spread and the cavity ohmic loss were not taken into consideration.

Table 3. Initial operating parameters of the electron beam.

Beam Voltage V_0	Magnetic Field B_0	Beam Current I_b	Pitch Factor α	Beam Radius R_b
40 kV	8.40 T	5 A	1.1	1.09 mm

Under the initial beam parameters, the simulation results of the rigorously confocal cavity, where $L_\perp = R_c = 4.20$ mm, are illustrated in Figure 5. As can be seen, the confocal cavity could generate a stable output power of 21.2 kW at a single frequency of 222.7 GHz, corresponding to an interaction efficiency of 10.6%. The field distributions shown in Figure 5c,d predict that the operating mode is a TE_{061} mode, as expected.

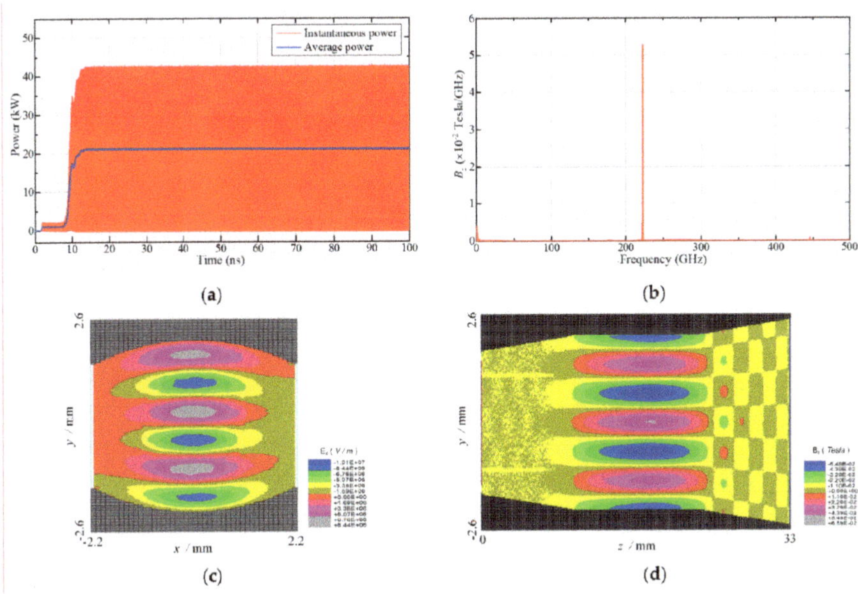

Figure 5. Typical simulation results of a rigorously confocal cavity: (**a**) instantaneous and average output power; (**b**) spectrum of the output magnetic field component B_z; (**c**) transverse distribution of the electric field component E_x; (**d**) axial distribution of the magnetic field component B_z.

3.1. Magnetic Field Tuning for Confocal Cavity

The frequency tuning characteristic of the proposed quasi-optical gyrotron cavity is first studied by changing the operating magnetic field, as the tuning method for traditional gyrotrons does. For the mirror separation $L_\perp = R_c = 4.20$ mm in a rigorously confocal cavity, the external magnetic field B_0 varying from 8.26 to 9.02 T is simulated when the other beam parameters are the same as the initial setting listed in Table 3. The oscillation frequency and the output power are as shown in Figure 6. As B_0 increases from 8.30 to 8.98 T, the oscillation frequency increases from 222.55 to 226.15 GHz with a frequency tuning range of 3.6 GHz, while the output power changes between 37.2 kW (at 8.32 T) and 1.1 kW (at 8.58 T). On the other hand, the excited transverse mode is the TE$_{15}$ mode for B_0 at less than 8.30 T, and the TE$_{16}$ mode for B_0 at larger than 8.98 T, rather than the expected TE$_{06}$ mode.

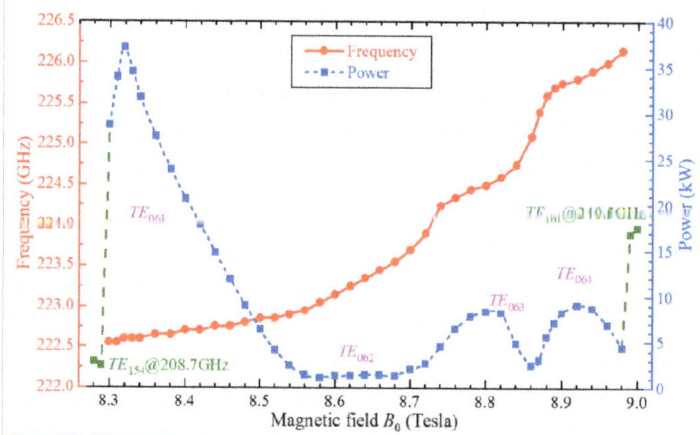

Figure 6. Simulation results of the oscillation frequency and the output power depending on the magnetic field B_0 for a confocal cavity.

The variation trends of both frequency and power in a confocal cavity are similar to those of the HOAMs excited in a long cylindrical cavity as reported in Reference [28]. In fact, the observation results of axial field distribution suggest that the first four axial modes of TE$_{06q}$ (q = 1, 2, 3, 4) are excited in the confocal cavity during the whole simulation. There are two obvious frequency jump points around 8.73 and 8.85 T. The reason for this is that the operating axial mode varies from one to another in these regions, somewhat affecting the continuity of frequency tuning. However, compared with the performances of conventional gyrotrons operating in this frequency band or output power level, a 3.6 GHz frequency tuning range has provided a great advantage, which should be attributed to the good mode-selective characteristics of a quasi-optical waveguide.

3.2. Mirror Separation Adjustment

Another kind of frequency tunability for the open quasi-optical cavity related to the mirror separation distance, as introduced in Section 2, is also investigated by PIC simulation.

By parallel shifting the upper and lower parts of the cavity in the y-direction, the separation distance between the two mirrors $L_\perp$ is adjusted from 4.10 to 4.60 mm. With the fixed operating parameters listed in Table 3, the simulation results for the oscillation frequency and output power are shown in Figure 7. At a fixed magnetic field B_0 = 8.4 T, with an increase in mirror separation distance $L_\perp$ from 4.13 mm to 4.56 mm, the oscillation frequency continuously decreases from 227.15 GHz to 210.25 GHz. The output power fluctuates between 29.5 kW (at 4.16 mm) and 1.3 kW (at 4.32 mm). Like in magnetic tuning, the spurious TE$_{15}$ and TE$_{16}$ modes are also excited at the lower and higher frequency edge, respectively.

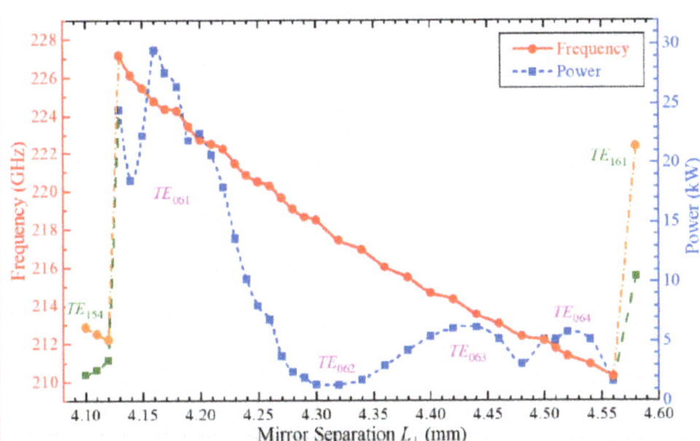

Figure 7. Simulation results of the oscillation frequency and the output power depending on the mirror separation $L_\perp$ at a fixed magnetic field 8.4 T.

Through adjusting mirror separation, the quasi-optical cavity covers a 17.1 GHz smooth frequency tuning band (about a relative bandwidth of 7.8% to the center frequency of 220 GHz), which is much broader than the 3.6 GHz acquired using a magnetic tuning mechanism. Although the continuous transitions of the operating axial mode from TE_{061} to TE_{064} have still been observed, the frequency tunability in mirror separation adjustment exhibits a much smoother tendency than that of the magnetic tuning shown in Figure 6.

3.3. Hybrid Tuning

As mentioned in Section 1, for the two items of the electron cyclotron frequency Ω_c and the EM wave frequency ω, there is only one variable that is alternated during the above simulations. Due to the mismatch of the cyclotron resonance conditions, the output powers are not stable and present a wide fluctuation in both frequency tuning manners. In this case, to maintain the output power at a relatively stable level, a hybrid tuning approach adjusting both the mirror separation $L_\perp$ and the magnetic field B_0 is simulated. During this simulation, under each separation distance $L_\perp$, the magnetic field value B_0 is scanned to find the peak output power while the other beam parameters are still as the values listed in Table 3.

The simulated dependencies of the peak output power and its corresponding oscillation frequency on the mirror separation distance $L_\perp$ are plotted in Figure 8, which also illustrates the corresponding values of the magnetic field B_0. It can be found that the output power is able to be kept at a high level, over 20 kW (an interaction efficiency of 10%), by adjusting the magnetic field correspondingly, while the maximum power is about 34.4 kW at $L_\perp = 4.22$ mm. The oscillation frequency covers an extraordinarily broad bandwidth of 41.55 GHz, from 245.50 to 203.95 GHz, when $L_\perp$ varies within ± 0.40 mm around the confocal distance $L_\perp = R_c = 4.20$ mm. By this tuning method, the frequency tuning bandwidth achieves 18.9% relative to the center frequency of 220 GHz. The frequency tuning characteristics predicted by PIC simulation are consistent with those of cold cavity analysis in Figure 4. Besides this, the axial field distribution results predict that the TE_{061} mode has always been the operating mode of every optimized parameter for hybrid tuning. That is why the oscillation frequency in Figure 8 varies almost linearly with the mirror separation distance. Therefore, the frequency tunability of hybrid tuning possesses great characteristics of smooth continuity, wide bandwidth, and high power.

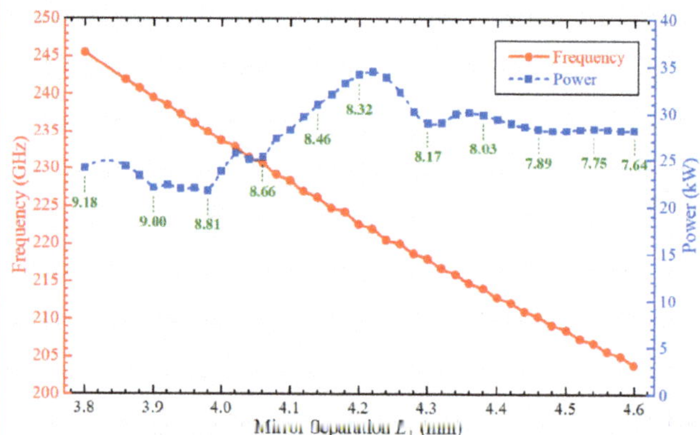

Figure 8. Simulation results of the peak output power and the corresponding oscillation frequency depending on the mirror separation $L_\perp$ during hybrid tuning. The green labels are the values of the corresponding magnetic field B_0 in Tesla.

The frequency tuning range is much wider than the requirements of high-power THz applications at present. In principle, it is possible to achieve a lower output frequency by further increasing the separation distance between two mirrors. PIC simulation result predicts that the quasi-optical cavity could still be operated in the TE_{061} mode if $L_\perp = 5.20$ mm, and generate an output power of 21 kW at 181.05 GHz.

4. Conclusions

In this paper, the frequency tuning characteristics of a high-power sub-THz quasi-optical gyrotron cavity based on a confocal waveguide have been investigated by PIC simulation. The possibility of smooth continuous frequency tuning over a broader band in a high-power sub-THz gyrotron is demonstrated by the simulation results. By continuously varying the mirror separation distance of a quasi-optical cavity and adjusting the operating magnetic field correspondingly, an extraordinarily wide frequency tuning band of 18.9% around 0.22 THz can be realized in a single mode of gyrotron operation, while the output power is kept at more than 20 kW. With the advantages of broad bandwidth, smooth tuning, and high power, this frequency tuning approach can be promoted to other bands and may be beneficial to the development of high-power frequency-tunable THz gyrotrons for some modern emerging applications.

Author Contributions: X.G. and W.F. contributed to the overall study design, analysis, computer simulation, and writing of the manuscript. J.Z., D.L., T.Y., Y.Y. and X.Y. provided technical support and revised the manuscript. All authors have read and agreed to the published version of the manuscript.

Funding: This work was supported in part by the National Key Research and Development Program of China No. 2019YFA0210202, in part by the National Natural Science Foundation of China under Grant 61971097 and Grant 6201101342, in part by the Sichuan Science and Technology Program No. 2018HH0136, and in part by the Terahertz Science and Technology Key Laboratory of Sichuan Province Foundation No. THZSC201801.

Data Availability Statement: Data sharing not applicable.

Conflicts of Interest: The authors declare no conflict of interest.

References

1. Chu, K.R. The Electron Cyclotron Maser. *Rev. Mod. Phys.* **2004**, *76*, 489–540. [CrossRef]
2. Nusinovich, G.S.; Thumm, M.; Petelin, M.I. The gyrotron at 50: Historical overview. *J. Infr. Millim. Terahertz Waves* **2014**, *35*, 325–381. [CrossRef]

3. Thumm, M. State-of-the-art of high-power gyro-devices and free electron masers. *J. Infr. Millim. Terahertz Waves* **2020**, *41*, 1–140. [CrossRef]
4. Rzesnicki, T.; Piosczyk, B.; Kern, S.; Illy, S.; Jin, J.B.; Samartsev, A.; Thumm, M. 2.2-MW record power of the 170-GHz European preprototype coaxial-cavity gyrotron for ITER. *IEEE Trans. Plasma Sci.* **2010**, *38*, 1141–1149. [CrossRef]
5. Glyavin, M.Y.; Denisov, G.G.; Zapevalov, V.E.; Koshelev, M.A.; Tretyakov, M.Y.; Tsvetkov, A.I. High power terahertz sources for spectroscopy and material diagnostics. *Phys. Uspekhi* **2016**, *59*, 595–604. [CrossRef]
6. Koshelev, M.A.; Tsvetkov, A.I.; Morozkin, M.V.; Glyavin, M.Y.; Tretyakov, M.Y. Molecular gas spectroscopy using radioacoustic detection and high-power coherent subterahertz radiation sources. *J. Mol. Spectrosc.* **2017**, *331*, 9–16. [CrossRef]
7. Nanni, E.A.; Barnes, A.B.; Griffin, R.G.; Temkin, R.J. THz dynamic nuclear polarization NMR. *IEEE Trans. Terahertz Sci. Technol.* **2011**, *1*, 145–163. [CrossRef]
8. Yamazaki, T.; Miyazaki, A.; Suehara, T.; Namba, T.; Asai, S.; Kobayashi, T.; Saito, H.; Ogawa, I.; Idehara, T.; Sabchevski, S. Direct observation of the hyperfine transition of ground-state positronium. *Phys. Rev. Lett.* **2012**, *108*, 253401. [CrossRef]
9. Nusinovich, G.S.; Luo, L.; Liu, P.K. Linear theory of frequency pulling in gyrotrons. *Phys. Plasmas* **2016**, *23*, 053111. [CrossRef]
10. Temkin, R.J. Development of terahertz gyrotrons for spectroscopy at MIT. *Terahertz Sci. Technol.* **2014**, *7*, 1–9. [CrossRef]
11. Blank, M.; Borchard, P.; Cauffman, S.; Felch, K.; Rosay, M.; Tometich, L. Development of high-frequency cw gyrotrons for DNP/NMR applications. *Terahertz Sci. Technol.* **2016**, *9*, 177–186. [CrossRef]
12. Idehara, T.; Tatematsu, Y.; Yamaguchi, Y.; Khutoryan, E.M.; Kuleshov, A.N.; Ueda, K.; Matsuki, Y.; Fujiwara, T. The Development of 460 GHz gyrotrons for 700 MHz DNP-NMR spectroscopy. *J. Infr. Millim. THz Waves* **2015**, *36*, 613–627. [CrossRef]
13. Liu, D.; Song, T.; Hu, Q.; Huang, J.; Zhang, Y.; Zhang, C.; Wang, W.; Hu, M. Detailed investigations on a multisection cavity for a continuously frequency-tunable gyrotron. *IEEE Trans. Electron Devices* **2019**, *66*, 2746–2751. [CrossRef]
14. Bratman, V.L.; Fedotov, A.E.; Kalynov, Y.K.; Osharin, I.V.; Zavolsky, N.A. Smooth wideband frequency tuning in low-voltage gyrotron with cathode-end power output. *IEEE Trans. Electron Devices* **2017**, *64*, 5147–5150. [CrossRef]
15. Chang, T.H.; Idehara, T.; Ogawa, I.; Agusu, L.; Kobayashi, S. Frequency tunable gyrotron using backward-wave components. *J. Appl. Phys.* **2009**, *105*, 063304. [CrossRef]
16. Guan, X.; Fu, W.; Lu, D.; Yan, Y.; Yang, T.; Yuan, X. Experiment of a High-Power Sub-THz Gyrotron Operating in High-Order Axial Modes. *IEEE Trans. Electron Devices* **2019**, *66*, 2752–2757. [CrossRef]
17. Glyavin, M.Y.; Luchinin, A.G.; Morozkin, M.V.; Khizhnyak, V.I. Smooth wideband tuning of the operating frequency of a gyrotron. *Radiophys. Quantum Electron.* **2008**, *51*, 57–63. [CrossRef]
18. Glyavin, M.Y.; Khizhnyak, V.I.; Luchinin, A.G.; Idehara, T.; Saito, T. The Design of the 394.6 GHz Continuously Tunable Coaxial Gyrotron for DNP Spectroscopy. *Int. J. Infrared Milli. Waves* **2008**, *29*, 641–648. [CrossRef]
19. Bratman, V.L.; Kalynov, Y.K.; Kalynova, G.I.; Manuilov, V.N.; Makhalov, P.B. Frequency tuning in a subterahertz gyrotron with a variable cavity. *IEEE Trans. Electron Devices* **2014**, *61*, 3529–3533. [CrossRef]
20. Fu, W.; Guan, X.; Yan, Y. Generating High-Power Continuous-Frequency Tunable Sub-Terahertz Radiation from a Quasi-Optical Gyrotron with Confocal Waveguide. *IEEE Electron Device Lett.* **2020**, *41*, 613–616. [CrossRef]
21. Alberti, S.; Tran, M.Q.; Hogge, J.P.; Tran, T.M.; Bondeson, A.; Muggli, P.; Perrenoud, A.; Jödicke, B.; Mathews, H.G. Experimental measurements on a 100 GHz frequency tunable quasioptical gyrotron. *Phys. Fluids B Plasma Phys.* **1990**, *2*, 1654–1661. [CrossRef]
22. Sirigiri, J.R.; Shapiro, M.A.; Temkin, R.J. High-power 140-GHz quasioptical gyrotron traveling-wave amplifier. *Phys. Rev. Lett.* **2003**, *90*, 258302. [CrossRef]
23. Hu, W.; Shapiro, M.; Kriescher, K.E.; Temkin, R.J. 140-GHz gyrotron experiments based on a confocal cavity. *IEEE Trans. Plasma Sci.* **1998**, *26*, 366–374. [CrossRef]
24. Guan, X.; Fu, W.; Yan, Y. A 0.4-THz second harmonic gyrotron with quasi-optical confocal cavity. *J. Infr. Millim. THz Waves* **2017**, *38*, 1457–1470. [CrossRef]
25. Guan, X.; Chen, C.; Fu, W.; Yan, Y.; Liu, S. Design of a 220-GHz continuous frequency-tunable gyrotron with quasi-optical cavity. In Proceedings of the 2015 IEEE International Vacuum Electronics Conference (IVEC), Beijing, China, 27–29 April 2015; pp. 1–2. [CrossRef]
26. Guan, X.; Fu, W.; Yan, Y. Continuously frequency-tunable 0.22 THz gyrotron oscillator with quasi-optical resonator. *Terahertz Sci. Technol.* **2016**, *9*, 166–176. [CrossRef]
27. Zhou, J.; Liu, D.; Liao, C.; Li, Z. CHIPIC: An efficient code for electromagnetic PIC modeling and simulation. *IEEE Trans. Plasma Sci.* **2009**, *37*, 2002–2011. [CrossRef]
28. Jawla, S.K.; Griffin, R.G.; Mastovsky, I.A.; Shapiro, M.A.; Temkin, R.J. Second Harmonic 527-GHz Gyrotron for DNP-NMR: Design and Experimental Results. *IEEE Trans. Electron Devices* **2020**, *67*, 328–334. [CrossRef]

electronics

Article

Study of 3D-Printed Dielectric Barrier Windows for Microwave Applications

Mikhail D. Proyavin , Dmitry I. Sobolev, Vladimir V. Parshin, Vladimir I. Belousov, Sergey V. Mishakin and Mikhail Y. Glyavin *

Institute of Applied Physics of the Russian Academy of Sciences, 603950 Nizhny Novgorod, Russia; pmd@ipfran.ru (M.D.P.); sobolev@ipfran.ru (D.I.S.); parsh@ipfran.ru (V.V.P.); vbelousov@ipfran.ru (V.I.B.); mishakin@ipfran.ru (S.V.M.)
* Correspondence: glyavin@ipfran.ru

Abstract: 3D printing technologies offer significant advantages over conventional manufacturing technologies for objects with complicated shapes. This technology provides the potential to easily manufacture barrier windows with a low reflection in a wide frequency band. Several 3D printing methods were examined for this purpose, and the dielectric properties of the various types of materials used for 3D printing were experimentally studied in the frequency range 26–190 GHz. These measurements show that the styrene-butadiene-styrene and polyamide plastics are suitable for broadband low-reflection windows for low-to-medium-power microwave applications. Two barrier windows with optimized surface shapes were printed and tested. Results demonstrate that the studied technique can fabricate windows with a reflection level below −18 dB in the frequency band up to 160 GHz. Studied windows can be used for spectroscopic tasks and other wideband microwave applications.

Keywords: dielectric properties; low-reflection barrier windows; broadband window; microwaves; terahertz radiation

check for **updates**

Citation: Proyavin, M.D.; Sobolev, D.I.; Parshin, V.V.; Belousov, V.I.; Mishakin, S.V.; Glyavin, M.Y. Study of 3D-Printed Dielectric Barrier Windows for Microwave Applications. *Electronics* **2021**, *10*, 2225. https://doi.org/10.3390/ electronics10182225

Academic Editor: Anna B. Piotrowska

Received: 30 June 2021
Accepted: 9 September 2021
Published: 10 September 2021

Publisher's Note: MDPI stays neutral with regard to jurisdictional claims in published maps and institutional affiliations.

1. Introduction

Additive manufacturing technologies have great potential for industry, science, and technology. There are a number of tasks that are difficult or almost impossible to implement with traditional fabrication methods. In particular, 3D printing from dielectric materials is a highly convenient and cheap tool for prototyping and manufacturing radiofrequency components. It creates a method of readily obtaining components with a sophisticated surface shape. For low-reflection microwave windows, a subwavelength grating with a specially designed shape at both sides of the window disk could significantly reduce the reflection coefficient of incident radiation in a wide frequency band [1]. This paper explores the possibility of using 3D printing to make millimeter-wave barrier windows that can operate in a wide frequency range.

Currently, there are several areas in which windows with the broadband transmission of low-power microwave radiation are used. These areas include molecular gas spectroscopy and DNP/NMR spectroscopy [2,3], measurements of fine positronium structure [4], which require both broadband-tunable microwave radiation sources such as backward-wave oscillators, gyrotrons, orotrons, and input windows in the working chamber of the spectrometer. Other examples are the radiometers and geophysical instruments used for atmospheric transparency studies, which require windows with a low reflection coefficient at different frequencies, corresponding to atmospheric transparency windows, with a bandwidth up to 20 GHz [5]. The cryogenic resonator complex requires an even wider frequency band [6], which needs a low reflection from the input and output windows in the entire operating frequency band (50–500 GHz). Reflections from windows lead to spurious interference, which ultimately reduces the sensitivity of the spectrometer. Such devices are currently equipped with lenses with concentric triangle-shaped grooves, which are noticeably worse in terms of reflections compared to the alternative surfaces considered in this paper [7]. At the same time, it is well known that reflections dramatically affect the gyrotron operation regime [8]. For all the mentioned applications, the radiation power does not exceed a few watts, which allows one to use windows made of polymers without the risk of overheating.

There are different methods to obtain the broadband transmission of microwave radiation through the windows. The almost reflectionless transmission of a linearly polarized wave can be achieved for the Brewster-angle disk; however, this output suffers from an inefficient use of space, which may be especially important for divergent wave beams and some distortion of the transverse field structure [9]. Polymers can be mixed with nanoparticles to produce a multilayer dielectric coating [10]. Metamaterial devices and gradient index photonic structures are also used to reduce reflection [11]. Another known method considered in this paper is the use of windows with a surface grating of a specially optimized shape, providing a significant reduction in reflection [12]. For the manufacture of such a surface, it is highly convenient to use 3D printing. Besides low cost and time consumption, 3D printing has no restrictions on the curvature of the surface compared to Computerized Numerical Control (CNC) machining. Furthermore, it has advantages in creating thin elements from relatively soft and brittle materials.

There are several different 3D printing technologies, each of which has its characteristics and uses its own types of polymers. Fused deposition modeling (FDM) technology [13] is readily available, easy to operate, and allows the use of a wide range of materials, including polyethylene, which has excellent properties in transmitting high-frequency radiation. However, this type of 3D printing is characterized by a low accuracy (up to 100 microns) and sufficiently pronounced layering, limiting the frequency range of applications. In addition, the print resolution in the transverse plane is limited by the diameter of the nozzle and the quality of the alignment. Selective laser sintering (SLS) technology [14], due to a similar process of plastic melting, also has a wide selection of materials. However, compared to FDM, it has higher accuracy and better resolution; a layer thickness of several tens of microns, a resolution in the transverse plane approximately equal to the size of the pellets of the plastic used. Finally, Photopolymer 3D printing (stereolithography, SLA) provides excellent print quality but currently proposes a limited choice of printing materials. In this paper, the SLS method was used for printing the studied window samples.

Currently, many different materials for 3D printing are presented on the market, and their number is constantly growing. Unfortunately, manufacturers usually do not provide information on the dielectric properties of materials, especially in the microwave, millimeter, and terahertz regions. In order to design the microwave components, the dielectric properties are of great importance, so the characterization of the materials used for additive manufacturing is needed.

This paper is organized as follows. In Section 2, the results of measurements of the dielectric properties of the commercially available plastics are described. In Section 3, the measurements of the transmission and reflection coefficients for 3D-printed windows with several surface shape profiles are presented, and the results are compared with conventionally manufactured windows. Section 4 discusses the transmitted power restrictions for the printed windows due to thermal properties. Finally, the results of the study are discussed and concluded in Section 5.

2. Characterization of Dielectric Properties of Plastics Materials for 3D Printing

To study the loss tangent and dielectric constant of the plastics, two independent methods were used to increase the reliability of the results. In the first method, a rod made of the investigated plastic was inserted into a rectangular waveguide. The reflection and transmission coefficients were measured, and the dielectric properties were calculated from the experimentally obtained frequency dependences. Formulas for reflection and transmission of the dielectric waveguide plug can be found in [15]. The scheme of the measurement is shown in Figure 1a. The measurements were made in the entire Ka frequency band (26–40 GHz), and the measured reflection and transmission curves were matched by analytic curves calculated using the constant dielectric permittivity approximation. It can be seen in Figure 1b that the constant dielectric permittivity approximation fits the measured data well.

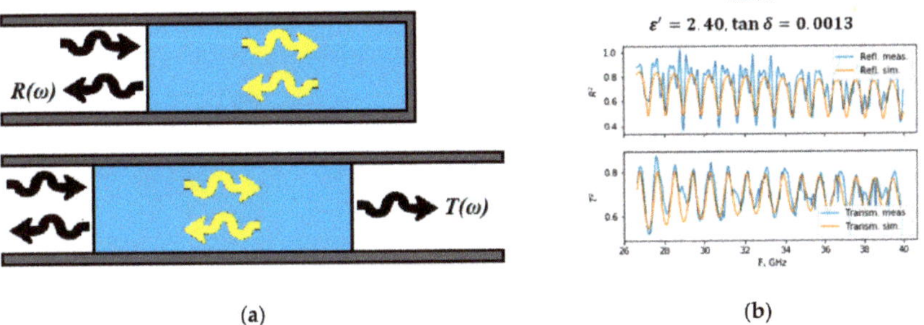

Figure 1. (**a**) The scheme of measurement of the reflection and transmission coefficients with the test sample in the form of the waveguide insert 7.2 mm × 3.4 mm × 100 mm (**a**); (**b**) comparison of the measurement results and simulation data for SBS plastic in the frequency range 26–40 GHz.

During the experiments, polymer samples printed using various technologies were studied. Results of dielectric permittivity and loss tangent measurements of various plastics printed using different 3D printing technologies at 100% infill are presented in Table 1. The dispersion of dielectric permittivity is relatively small within the Ka-band. Analysis of experimental data shows that styrene-butadiene-styrene (SBS) and polyethylene terephthalate glycol (PETG) are the most suitable materials for windows due to low losses and moderate dielectric permittivity. Different samples of PETG and SBS were printed using filaments from different manufacturers and showed minor differences. However, we note that these types of plastics were only suitable for FDM technology printers, which provided fair accuracy and thus were not applicable at sub-terahertz frequencies but sufficient for frequencies of several tens of GHz. This method and this plastic were used earlier, in particular, to print a two-dimensional Bragg resonator operating in the frequency range 55–65 GHz, and the measurement results were in good agreement with the theory [16]. Since the windows for spectroscopic and atmospheric measurements are also required at higher frequencies, the study of the applicability of FDM printing for these purposes is of interest.

Table 1. Dielectric properties of the 3D-printed samples in Ka-band.

Plastic	Printer	Re ε	tan δ
Polyethylene terephthalate glycol (PETG) 1	FDM	2.31	1.5×10^{-3}
PETG 2	FDM	2.47	1.6×10^{-3}
Polylactic Acid (PLA)	FDM	2.27	6.2×10^{-3}
Sterol-butadiene-sterol (SBS) 1	FDM	2.22	1.6×10^{-3}
SBS 2	FDM	2.40	1.3×10^{-3}
Visijet SL Clear	SLA	1.8	3.3×10^{-2}
Visijet SL Flex	SLA	1.8	3.5×10^{-2}
Visijet SL Hi-Temp	SLA	1.8	3.5×10^{-2}
Polyamide	SLS	1.7	3.5×10^{-3}

The second method for measuring the properties of dielectrics was to place samples in the form of flat disks between the mirrors of a high-quality open two-mirror Fabry-Perot resonator. By measuring the quality factor of an empty cavity and a cavity with a dielectric insert, it is possible to measure the properties of the test sample with high accuracy [17]. These measurements were made for selected materials and frequencies up to 185 GHz. The real part of the dielectric permittivity is the same as in the Ka-band measurements. The scheme of the measurement setup is shown in Figure 2a. The loss tangent data for plastic used for FDM printing (SBS) and plastic used for SLS printing (polyamide) is shown in Figure 2b.

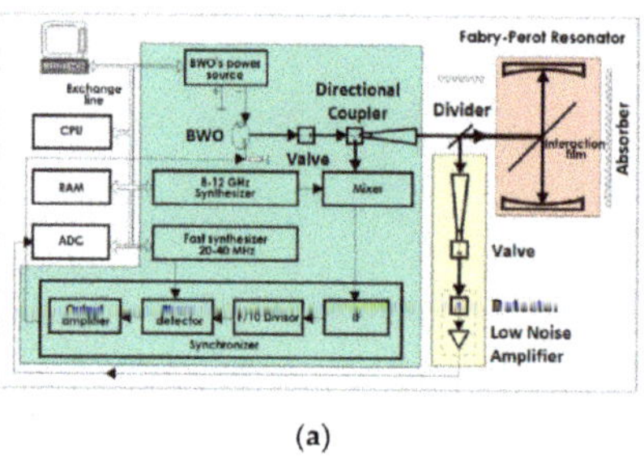
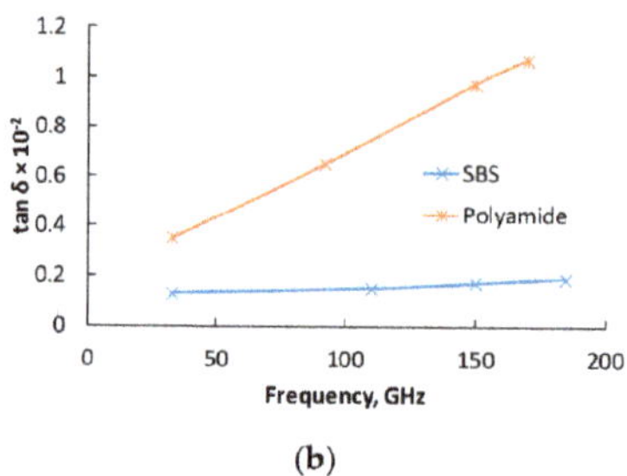

(a)   (b)

Figure 2. (a) The scheme of measurement of dielectric properties with the test sample in the form of the disk in the quasioptical resonator; (b) loss tangent of the SBS and polyamide.

The polyamide plastic for SLS printers has a somewhat higher absorption than SBS and PETG. Therefore, it could be an optimal solution for some applications as a trade-off between the higher losses and better printing quality of the SLS method.

The photopolymer materials have significantly larger losses than polyamide (tangent delta higher than 0.03). However, we consider these materials a good solution for high-frequency applications with a 1 milliwatt or lower level of microwave power, since SLA printing allows much better print accuracy and surface quality.

3. Reflection Measurements for 3D-Printed Prototype Barrier Windows

For the experimental study of the broadband windows prototypes, we chose SLS printing from polyamide due to low losses and high manufacturing accuracy, which allow the creation of small-scale structures suitable for devices operating at frequencies of several hundred GHz. The sizes of the subwavelength antireflection structures were chosen to produce fine details by the selected printing method adequately. These structures should perform well when half of the wavelength is bigger than the period of the structure because there could be no ±1st order diffraction scattering in these conditions. However, the performance at higher frequencies might degrade faster or slower depending on the shape of the elements. This paper considers the two variants of known antireflection subwavelength gratings at window disks for additive manufacturing. The first variant of the surface shape is a periodic array of pyramids with a base size smaller than the wavelength [18]. The shape of the pyramidal grating is shown in Figure 3a. The advantages of such a surface are a weak dependence of the reflection coefficient on the frequency and polarization of the incident radiation. The base side of the pyramids was chosen to be 1 mm, and the height was 2 mm. Simulations show that this corrugation applied to both surfaces of the polyamide window disk and provided reflections of less than −20 dB in the frequency band wider than one octave. For comparison, the flat polyamide disk had reflections of up to −9 dB in this band. The second option is a one-dimensional periodic corrugation of a special shape, proposed in [19]. This profile is optimized to minimize the reflection of the polarization with the direction of the electric field orthogonal to the groove direction at the frequency range of 60–160 GHz. The shape of the grooves is shown in Figure 3b. The period and depth of one-dimensional corrugation are 2 mm and 2.5 mm, respectively. The corrugation profile was optimized for one linear polarization only, and the reflection coefficient was less than -20 dB for frequencies below 105 GHz.

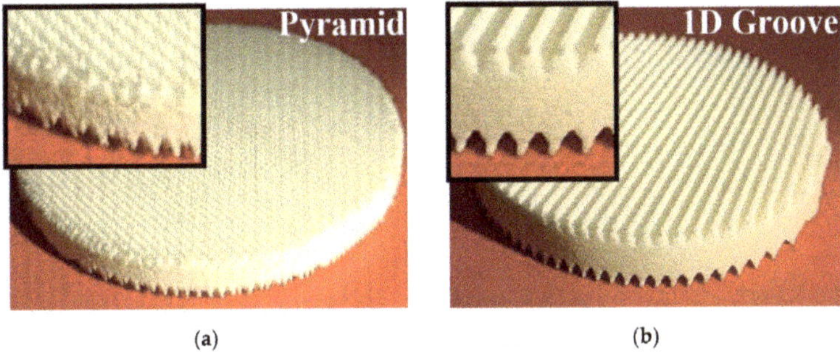

(a)   (b)

Figure 3. 3D-printed windows with (**a**) pyramids; (**b**) one-dimensional grating of special shape.

Three discs were printed: the first one had a flat surface on both sides, the other two had gratings of the tested shape on both sides. The reflection coefficients were obtained by two measurement setups by a vector network analyzer with a step of 30 MHz in the bands of 75–110 GHz and 130–160 GHz, where the disks with optimized surface shapes calculated reflection minima. The window disks were attached to the output of the corrugated tapers providing the gaussian wave beam flat phase front (Figure 4a). The 0-dB level of the setup was set using the flat mirror closing the end of the taper, and the minimum sensitivity limit was set as the reflection from the open end of the taper.

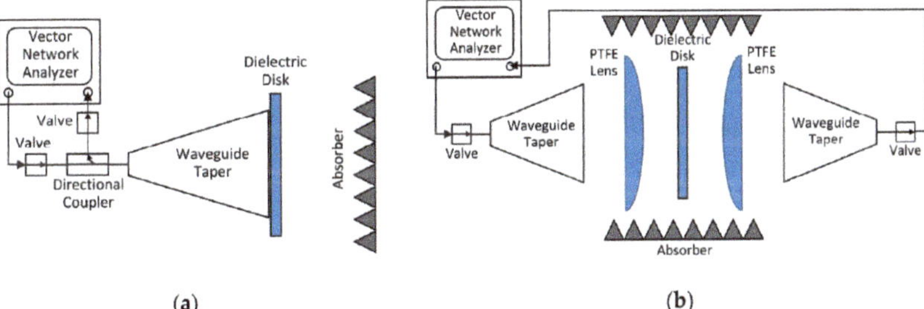

(a)   (b)

Figure 4. Schematics of the setups used to measure the dielectric disks parameters: (**a**) reflection measurements; (**b**) transmission measurements.

Reflection measurement results are shown in Figures 5–7 for the flat-surface disk, disk with pyramids, and disk with one-dimensional corrugation, correspondingly. The disk with a one-dimensional corrugation of the surface was measured for both orthogonal linear polarizations. However, the results for the second polarization are significantly worse than for the optimal polarization. The measured reflection coefficient is below -18 dB for both corrugated disks in the frequency ranges of 75–110 GHz and 130–160 GHz. In contrast, the flat disk has a much narrower band with low reflections (10 GHz at the level of -20 dB). Note that the option with pyramids is more advantageous if radiation of arbitrary polarization is required, while the option with one-dimensional corrugation is better for specific linear polarization. The one-dimensional corrugation works better for lower frequencies but (due to its larger period) is worse for higher frequencies. Due to the high sensitivity of this shape to manufacturing tolerances, the measured reflection of one-dimensional corrugation significantly deviates from the calculated one, which is also noted in [19].

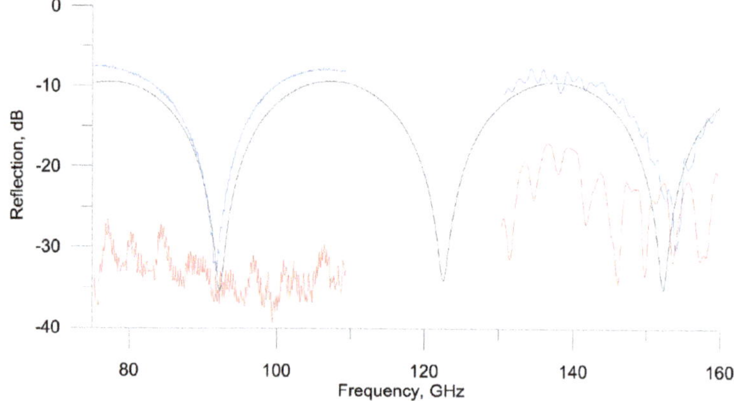

Figure 5. Reflection from the 3D-printed disk with a flat surface on both sides. The black line corresponds to the numerical simulation in CST Studio. The blue line is the measured reflection coefficient, and the red line is the lower sensitivity limit of the measurement setup.

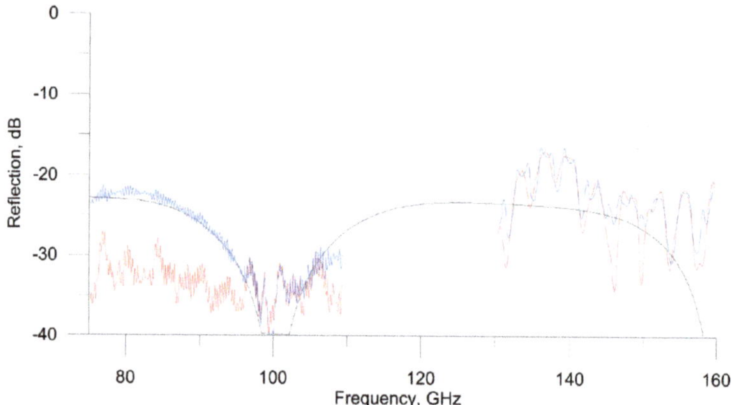

Figure 6. Reflection from the 3D-printed disk with pyramids on both sides. The black line corresponds to the numerical simulation in CST Studio. The blue line is the measured reflection coefficient, and the red line is the lower sensitivity limit of the measurement setup.

To measure wideband transmission, we used a quasioptical setup consisting of a vector network analyzer (VNA), a pair of tapers with PTFE lenses on adaptors, and an air gap between them (Figure 4b). Disks were placed in the center of the air gap, which is also the position of the Gaussian beam waist. The transmission through the two thinner disks is presented in Figure 8. The flat disk has a thickness of 3.15 mm, and the average thickness of the disk with pyramids is 3.33 mm. The low-reflection disk has a significantly better transmission and is very close to the maximum transmission predicted using the measured loss tangent of the polyamide. The disk with one-dimensional corrugation is not shown because it has high dielectric losses in this frequency band due to its bigger average thickness of 7.5 mm. The oscillations of the flat-sided disk transmission are caused by Fabry–Perot resonances inside the disk (period approximately 30 GHz) and resonances between the disk and one of the lenses (approximately 4 GHz). The reflection from the tapers causes fast oscillations (approximately 1 GHz).

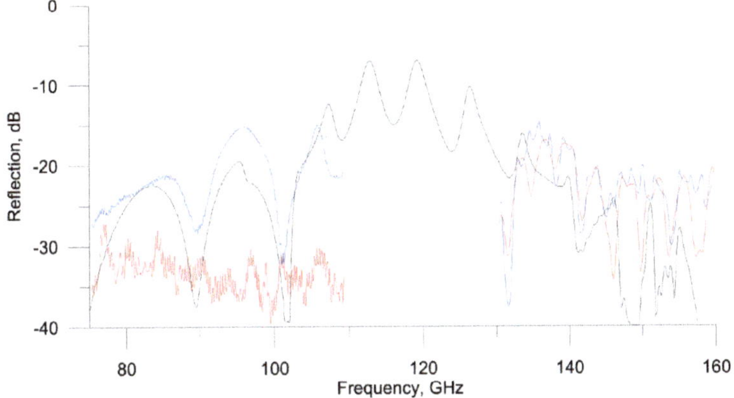

Figure 7. Reflection from the 3D-printed disk with special one-dimensional corrugation on both sides. The black line corresponds to the numerical simulation in CST Studio. The blue line is the measured reflection coefficient, and the red line is the lower sensitivity limit of the measurement setup.

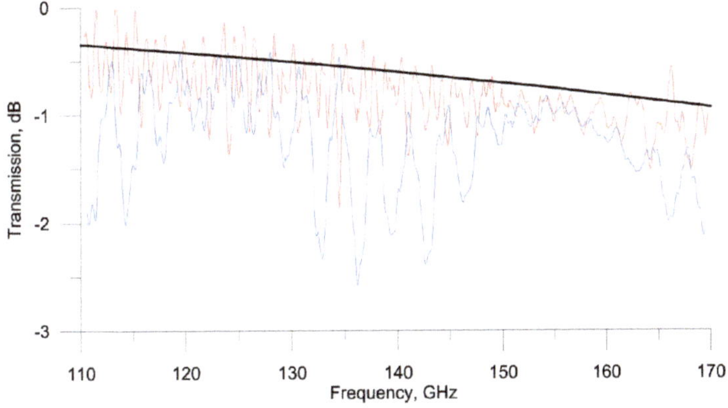

Figure 8. Transmission through the disks. The red line corresponds to the disk with pyramids on both sides, and the blue line corresponds to the flat-surface disk. The black line shows dielectric losses calculated in the 3.33 mm thick polyamide disk with zero reflections.

It is interesting and instructive to compare the results obtained for 3D-printed windows with the earlier results for windows with the same shape profiles but manufactured by traditional mechanical methods. Thus, the disk made from raflon (radiation-modified PTFE) by CNC machining had reflections lower than −20 dB in the frequency range of 50–200 GHz [19], which is very similar to the results presented in this paper. Therefore, we conclude that the current additive technologies are competitive with traditional manufacturing methods for dielectric microwave components at the 1 mm and longer waves.

4. Estimation of the Applicability of the Considered Materials

Based on the obtained values of the real and imaginary parts of the dielectric permittivity, as well as the mechanical and thermal parameters of the tested materials given in [20], it is possible to determine the losses and temperature conditions of a window that can withstand atmospheric pressure, depending on its diameter and the supplied microwave power. When considering the heat problem, the transverse dependence of the microwave radiation intensity on the radius was taken as a Gaussian wave beam with a width of 0.64 of the window radius. Numerical modeling was performed with the following parameters to represent many plastics with similar properties: $n = 1.5$, emissivity $\varepsilon = 0.9$, and the thermal conductivity of plastic $\kappa = 0.4$ W·m^{-1}K^{-1}. The dielectric loss tangent was taken as slightly larger than the best-tested plastic $\delta = 0.002$. The temperature of the cooled edge of the window (ambient temperature) was $T_0 = 20$ °C, with a convection coefficient of h $= 10$ W·m^{-2}·K^{-1}. Several window diameters were considered between 1 cm and 10 cm with different thicknesses from 1 mm to 10 mm, and the wavelength was set as $\lambda_0 = 3$ mm, which corresponded to the 100 GHz base frequency. Dielectric losses are proportional to the frequency; therefore, the maximum power for any other frequency can be calculated by multiplying the ratio of the base frequency to the target frequency. Numerical modeling of the dielectric disk heating by a Gaussian beam was performed in COMSOL Multiphysics. The reflections of the beam on disk surfaces were neglected, assuming the antireflection surfaces. We simulated maximum beam power, which can be transmitted through the disk, given that the maximum stationary temperature is 120 °C. The decimal logarithm of the maximum transmitted continuous wave (CW) power value in Watts is presented in Figure 9, e.g., a disk with diameter 100 mm and thickness 10 mm can withstand approximately 150 W of CW transmitted power at a frequency of 100 GHz without convective cooling.

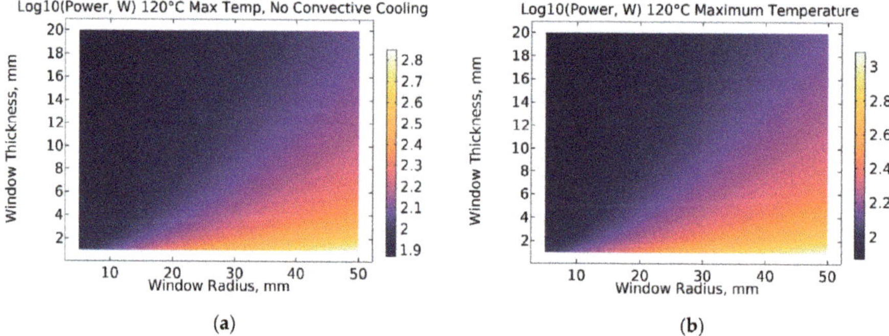

Figure 9. Simulation results for maximum 100 GHz CW transmitted power through the polymer windows without overheating. (a) the decimal logarithm of the maximum power value in watts without convective cooling (vacuum at both sides); (b) decimal logarithm of the maximum power value in watts with convective air cooling on one side.

5. Conclusions

Current 3D printing technologies allow easy-to-manufacture dielectric microwave components, such as broadband barrier windows with low reflections. In this paper, the dielectric properties of the materials used in various 3D printing methods were measured. The most suitable materials were found to be useful in microwave systems with a frequency of up to several hundred gigahertz and a power of up to several tens of watts. The analytical calculations and numerical simulations verify the use of the studied materials for microwave devices with a power of about 100–200 watts.

Barrier windows with surface shapes specially optimized for low reflection in a wide frequency band were printed and examined at low power. The printed windows provided a reflection coefficient below -18 dB at frequencies of up to 160 GHz, making it possible to use in microwave devices that require the reception/transmission of a signal in a wide frequency range. We conclude that the current additive technologies are competitive with traditional manufacturing methods for dielectric microwave components at 1 mm and for longer waves.

Author Contributions: Conceptualization, M.Y.G., M.D.P. and D.I.S.; methodology, M.D.P. and D.I.S.; investigation, M.D.P., D.I.S., V.V.P., V.I.B. and S.V.M.; writing—original draft preparation, M.D.P.; writing—review and editing, D.I.S.; project administration, M.Y.G. All authors have read and agreed to the published version of the manuscript.

Funding: This research was funded by Russian Science Foundation, grant number 21-19-00877.

Data Availability Statement: Data available on request from the authors.

Conflicts of Interest: The authors declare no conflict of interest.

References

1. Mirotznik, M.S.; Good, B.L.; Ransom, P.; Wikner, D.; Mait, J.N. Broadband antireflective properties of inverse motheye surfaces. *IEEE Trans. Antennas Propag.* **2010**, *58*, 2969–2980. [CrossRef]
2. Idehara, T.; Mitsudo, S.; Sabchevski, S.; Glyavin, M.Y.; Ogawa, I. Gyrotron FU series—Current status of development and applications. *Vacuum* **2001**, *62*, 123–132. [CrossRef]
3. Idehara, T.; Sabchevski, S.P.; Glyavin, M.; Mitsudo, S. The gyrotrons as promising radiation sources for THz sensing and imaging. *Appl. Sci.* **2020**, *10*, 980. [CrossRef]
4. Fedotov, A.E.; Rozental, R.M.; Zotova, I.V.; Ginzburg, N.S.; Sergeev, A.S.; Tarakanov, V.P.; Glyavin, M.Y.; Idehara, T. Frequency tunable sub-THz gyrotron for direct measurements of positronium hyperfine structure. *J. Infrared Millim. Terahertz Waves* **2018**, *39*, 975–983. [CrossRef]
5. Nosov, V.I.; Bolshakov, O.S.; Bubnov, G.M.; Vdovin, V.F.; Zinchenko, I.I.; Marukhno, A.S.; Nikiforov, P.L.; Fedoseev, L.I.; Shvetsov, A.A. A dual-wave atmosphere transparency radiometer of the millimeter-wave range. *Instrum. Exp. Tech.* **2016**, *59*, 374–380. [CrossRef]
6. Parshin, V.V.; Serov, E.A.; Bubnov, G.M.; Vdovin, V.F.; Koshelev, M.A.; Tretyakov, M.Y. Cryogenic resonator complex. *Radiophys. Quantum Electron.* **2014**, *56*, 554–560. [CrossRef]
7. Lamb, J.W. Cross-Polarisation and astigmatism in matching grooves. *Int. J. Infrared Millim. Waves* **1996**, *17*, 2159–2165. [CrossRef]
8. Glyavin, M.Y.; Zapevalov, V.E. Reflections influence on the gyrotron oscillation regimes. *Int. J. Infrared Millim. Waves* **1998**, *19*, 1499–1511. [CrossRef]
9. Li, Q.; Vernon, R.J. Reflection and transmission of a Gaussian beam from a dielectric window at the Brewster angle. In Proceedings of the 2005 Joint 30th International Conference on Infrared and Millimeter Waves and 13th International Conference on Terahertz Electronics, Williamsburg, VA, USA, 19–23 September 2005; Volume 2, pp. 513–514.
10. Cai, B.; Chen, H.; Xu, G.; Zhao, H.; Sugihara, O. Ultra-Broadband THz antireflective coating with polymer composites. *Polymers* **2017**, *9*, 574. [CrossRef] [PubMed]
11. Chen, Y.W.; Zhang, X.C. Anti-Reflection implementations for terahertz waves. *Front. Optoelectron.* **2014**, *7*, 243–262. [CrossRef]
12. Bräuer, R.; Bryngdahl, O. Design of antireflection gratings with approximate and rigorous methods. *Appl. Opt.* **1994**, *33*, 7875–7882. [CrossRef] [PubMed]
13. 3D Printing & Additive Manufacturing. Available online: https://www.stratasys.com/ (accessed on 29 June 2021).
14. Beaman, J.J.; Deckard, C.R. Selective Laser Sintering with Assisted Powder Handling. U.S. Patent 4938816A, 3 July 1990.
15. Rinkevich, A.B.; Perov, D.V.; Ryabkov, Y.I. Transmission, reflection and dissipation of microwaves in magnetic composites with nanocrystalline finemet-type flakes. *Materials* **2021**, *14*, 3499. [CrossRef] [PubMed]
16. Ginzburg, N.S.; Peskov, N.Y.; Zaslavsky, V.Y.; Kocharovskaya, E.R.; Malkin, A.M.; Sergeev, A.S.; Baryshev, V.R.; Proyavin, M.D.; Sobolev, D.I. 2D Bragg resonators based on planar dielectric waveguides (from theory to model-based testing). *Semiconductors* **2019**, *53*, 1282–1286. [CrossRef]
17. Vlasov, S.N.; Koposova, E.V.; Mazur, A.B.; Parshin, V.V. On permittivity measurement by a resonance method. *Radiophys. Quantum Electron.* **1996**, *39*, 410–415. [CrossRef]
18. Ma, J.Y.L.; Robinson, L.C. Night moth eye window for the millimetre and sub-millimetre wave region. *Opt. Acta Int. J. Opt.* **1983**, *30*, 1685–1695. [CrossRef]
19. Vlasov, S.N.; Koposova, E.V.; Kornishin, S.Y. Wideband windows for millimeter- and submillimeter-wave vacuum devices. *Radiophys. Quantum Electron.* **2020**, *63*, 115–123. [CrossRef]
20. Laureto, J.; Tomasi, J.; King, J.A.; Pearce, J.M. Thermal properties of 3-D printed polylactic acid-metal composites. *Prog. Addit. Manuf.* **2017**, *2*, 57–71. [CrossRef]

electronics

Article

Over-Size Pill-Box Window for Sub-Terahertz Vacuum Electronic Devices

Tongbin Yang [1], Xiaotong Guan [2,3], Wenjie Fu [1,3,*], Dun Lu [1], Chaoyang Zhang [1], Jie Xie [1], Xuesong Yuan [1,3] and Yang Yan [1,3]

1   School of Electronic Science and Engineering, University of Electronic Science and Technology of China, Chengdu 610054, China; yangtongbin@std.uestc.edu.cn (T.Y.); ludun@std.uestc.edu.cn (D.L.); chaoyangzhang@std.uestc.edu.cn (C.Z.); xiejie@std.uestc.edu.cn (J.X.); yuanxs@uestc.edu.cn (X.Y.); yanyang@uestc.edu.cn (Y.Y.)
2   School of Physics, University of Electronic Science and Technology of China, Chengdu 610054, China; guanxt@uestc.edu.cn
3   Terahertz Science and Technology Key Laboratory of Sichuan Province, University of Electronic Science and Technology of China, Chengdu 610054, China
*   Correspondence: fuwenjie@uestc.edu.cn

Abstract: The pill-box window is one of the important components of microwave vacuum electronic devices (VEDs), and research into it is of great significance. As the operating frequency increases, the problems associated with the reduction in the structure size include the reduction of the brazing plane and the reduction in the tolerance of the pill-box window. These problems will cause traditional pill-box windows to be unsuitable in high-frequency bands, especially in terahertz and sub-terahertz regions. The most influential factor is the length of the circular waveguide in the box window. The welding plane of the over-size pill-box window is the annular bottom surface on both sides of the dielectric sheet, which is larger than the circular waveguide, and the operating frequency does not directly affect the area of the brazing surface. Choosing a suitable diameter for the dielectric sheet can effectively increase the tolerance to the length of the pill-box window circular waveguide. Therefore, an over-size pill-box window would be a practicable approach to improve the performance compared to the traditional pillow-box in high-frequency bands. This paper describes, in detail, the theoretical design, simulation optimization and experimental process of this improved pill-box window. An over-size pill-box window suitable for G band VEDs was successfully developed. The experimental result in the 215–225 GHz band is that the maximum transmission loss is −1 dB, and the overall transmission loss is close to −0.5 dB. The overall reflection is less than −11 dB.

Keywords: pillbox window; 220 GHz; low loss; sub-terahertz

Citation: Yang, T.; Guan, X.; Fu, W.; Lu, D.; Zhang, C.; Xie, J.; Yuan, X.; Yan, Y. Over-Size Pill-Box Window for Sub-Terahertz Vacuum Electronic Devices. *Electronics* **2021**, *10*, 653. https://doi.org/10.3390/electronics10060653

Academic Editor: Mikhail Glyavin

Received: 8 February 2021
Accepted: 9 March 2021
Published: 11 March 2021

Publisher's Note: MDPI stays neutral with regard to jurisdictional claims in published maps and institutional affiliations.

1. Introduction

In Recent decades, with the rapid development of terahertz science and technology, the lack of terahertz radiation sources has become, have limiting the research into and applications of the terahertz wave. Vacuum electronic devices (VEDs) (e.g., TWT, klystron, BWO, Gyrotron), which could transfer energies from electrons to electromagnetic waves in vacuum tubes, are the most powerful radiation sources in the low-frequency terahertz region and sub-terahertz region. The IAP-RAS (Institute of Applied Electronics-RAS) reported 10 kW, 1 THz gyrotron, CAEP (China Academy of Engineering Physics) reported 3.1 W, 336.96 GHz TWT, and CPI (Communications & Power Industries) reported 10 W, 264 GHz EIO/EIK (Extended Interaction Klystrons/Oscillatiors). The vacuum window is an important component of the vacuum electronic device [1–3]. It is used to maintain the high-vacuum condition while in the tubes, while inputting or outputting electromagnetic waves from or into the tubes.

The pill-box window is a widely used configuration in VEDs design [4–7], with the advantages of bandwidth, easy brazing, and high power capacity. Figure 1a shows the standard structure of a traditional pill-box window. It consists of a standard waveguide with input and output ports, a circular waveguide with a diameter equal to the diagonal of the standard waveguide, and a dielectric sheet brazed into the circular waveguide. The traditional pill-box window has a wide range of applications. However, the size of the traditional pill-box window is directly proportional to the operating wavelength. As the working frequency increases, the overall size of the pill-box window is reduced, and some problems arise in the realization of the traditional pill-box windows a result. The brazing surface of the traditional pill-box window is the cylindrical side surface of the dielectric sheet. The reduction in the size of the pill-box window will cause the reduction in the brazing surface, decreasing the air tightness and experimental tolerances. The higher the working frequency band of the pill-box window, the more difficult it is to process. The parameter that is most prone to experimental errors, and has the greatest impact on the experimental results, is the length of the circular waveguide of the pill-box window (L_1) [5,6]. In the W-band, an asymmetric structure is used, and the influence of processing errors on the test results of the box window can be reduced through multiple transmission tests [6]. However, in the G-band, the asymmetric structure cannot solve the problem of a too-small welding surface, and the air tightness of the pill-box window cannot be guaranteed. The improved over-sized pill-box window shown in Figure 1b can effectively solve the above problems. The brazing surface of the improved pill-box window is the annular bottom surface of the dielectric sheet, which is larger than the circular waveguide. The brazing area will not change with the increase in working frequency. During the design process, we found that the tolerances in dielectric sheets with different diameters to pill-box windows are different. Choosing a suitable diameter for the dielectric sheet can effectively increase the tolerance of the pill-box window, thereby reducing the difficulty of achieving high-frequency pill-box windows.

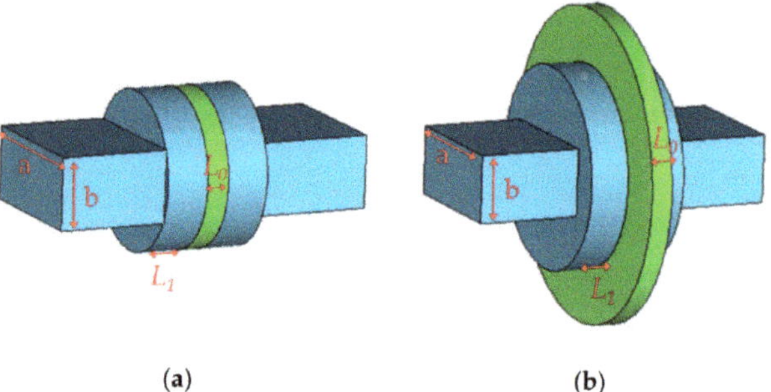

Figure 1. Structure of the (**a**) traditional pillbox window and (**b**) improved pillbox window.

In this paper, an over-size pill-box window for sub-terahertz VEDs at G band (215–225 GHz) is investigated, as outlined in the above structure. In the second part of this article, the theoretical design and simulation optimization of the improved box window are described in detail. The third part realized the G-band pill-box window according to the design and processing. The fourth part summarizes this article.

2. Model Design and Simulation

The pill-box window design begins with choosing the right material as the medium window sheet. Materials such as quartz, ceramics, diamond and sapphire are often used for the dielectric sheets of pill-box windows. Compared with other dielectric materials, sapphire has the advantages of high mechanical strength, low loss tangent, and mature metallization technology. Sapphire is an anisotropic material, but the perturbation of its dielectric constant does not affect the transmission of the transverse electric (TE) mode in the over-size pill-box window [7]. Therefore, the dielectric material selected in this design is sapphire, and its dielectric constant is 9.4.

There are many published theoretical calculation methods, such as the impedance-matching approach [3,5], the method of moments [8], and the equivalent circuit method [4–9]. In this paper, the equivalent circuit method, with a relatively simple calculation, is selected. In the equivalent circuit theory, the connection point between the waveguides and the waveguide of the pill-box window are equivalent to a two-port circuit element. The transmission modes of the rectangular waveguide and the circular waveguide in the pill-box window are TE_{10} mode and TE_{11} mode, and the equivalent circuit is shown in Figure 2.

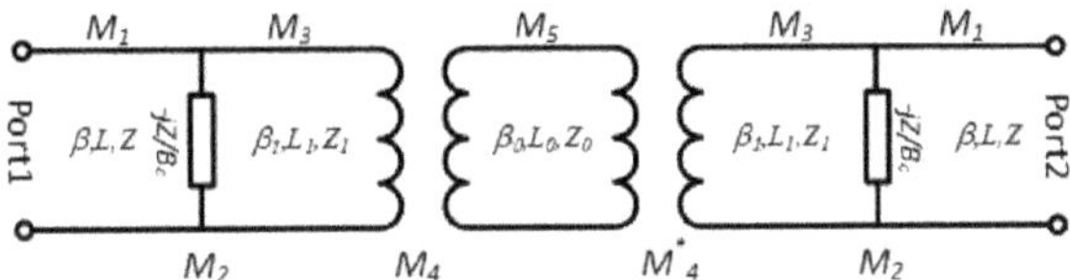

Figure 2. Equivalent circuit diagram of pill-box window.

In Figure 2, L is the length of the rectangular waveguide; β is the propagation constant of the rectangular waveguide; Z is the characteristic impedance of the rectangular waveguide; Bc is the equivalent susceptance of the connecting part of the rectangular waveguide and the circular waveguide; L_1 is the length of the cylindrical waveguide; β_1 is the propagation constant of the cylindrical waveguide; Z_1 is the characteristic impedance of the cylindrical waveguide; L_0 is the length of the sapphire cylinder; β_0 is the propagation constant of sapphire dielectric waveguide; Z_0 is the characteristic impedance of sapphire dielectric waveguide; a M is the transfer matrix of the connection part and each waveguide.

The multiplication of the transmission matrix of each the transition part and each waveguide of the pill-box window is equal to the overall transmission matrix of the pill-box window. This can be written as

$$M = M_1 \bullet M_2 \bullet M_3 \bullet M_4 \bullet M_5 \bullet M_4^* \bullet M_3 \bullet M_2 \bullet M_1 \tag{1}$$

where the '$\bullet$' in the formula means multiply, the '*' in the formula represents the inverse matrix. The specific expression formula of each transfer matrix can be found in the literature [7]. Through the simplification of the transfer matrix and the relationship between the scattering matrix and the transfer matrix, the scattering matrix of the equivalent circuit can be obtained. The specific derivation process can be obtained from the literature [6,7]. By substituting the ideal transmission conditions: $S_{12} = S_{21}$, $S_{11} = S_{22} = 0$ into the scattering matrix the equations for the structural parameters of the box window can be obtained as

$$\begin{cases} \tan \beta_1 L_1 = A/B \pm \sqrt{(A/B)^2 - C/B} \\ (A/B)^2 - C/B \geq 0 \\ A = \left(-\frac{Z}{Z_1} + \frac{Z_1}{Z}B_c^2 + \frac{Z}{Z_1}\right)\cos\beta_0 L_0 + \left(\frac{Z_1}{Z_0}B_c^2 + \frac{Z_0}{Z_1}B_c\right)\sin\beta_0 L_0 \\ B = -2B_c \cos\beta_0 L_0 + \left(\frac{Z_1^2}{ZZ_0}B_c^2 - \frac{ZZ_0}{Z_1^2} + \frac{Z_1^2}{ZZ_0}\right)\sin\beta_0 L_0 \\ C = 2B_c \cos\beta_0 L_0 + \left(\frac{Z}{Z_0} - \frac{Z_0}{Z}B_c^2 - \frac{Z_0}{Z}\right)\sin\beta_0 L_0 \end{cases} \tag{2}$$

According to the test conditions of the vector network analyzer (VNA) in our laboratory and the operating frequency of the pill-box window, the rectangular waveguide selected in this design is WR-5 (waveguide name of EIA standard). Equation (2) is an equation about f, L_1, L_0, R_1, R_0. When the frequency is determined to be 220 GHz, these four parameters have countless solutions. In order to obtain the ideal over-size pill-box window structure parameters, it is necessary to limit the values of the parameters before solving. The increase in the thickness of the sapphire sheet will increase the mechanical strength and power capacity of the pill-box window. However, the increase in thickness will reduce the matching degree of the pill-box window, that is, the bandwidth will become smaller. Therefore, the thickness of the sapphire sheet can be limited one-quarter waveguide wavelength to half the waveguide wavelength (0.11 mm $< L_0 <$ 0.22 mm). If the radius of the circular waveguide is too large, it is easy to produce higher-order modes. However, when the diameter of the circular waveguide is the diagonal length of the rectangular waveguide, the bandwidth of the box window is narrow. Therefore, the radius of the circular waveguide can be selected as $\sqrt{a^2 + b^2}/2 < R_1 < 1$ mm. To simplify the theoretical calculation, the radius of the dielectric sheet is temporarily set as $R_0 = R_1 + 1$. When the length of the circular waveguide is less than the length of the dielectric sheet, the matching degree of the box window will be reduced, and the bandwidth will be reduced. In order to reduce the calculation time, the length of the circular waveguide should be controlled within one waveguide wavelength. The length of the circular waveguide is between the length of the dielectric sheet and a waveguide wavelength ($L_0 < L_1 <$ 0.44 mm). Substituting the solutions in Equation (2) that meet the value ranges of the four parameters into the scattering matrix of the window system, take the solutions with the largest bandwidth with a reflection less than -15 dB as the theoretical design parameters of the pill-box window, as shown in Table 1.

Table 1. Structural size parameters of the pill-box windows.

	Equivalent Circuit	Simulation
Rectangular waveguide	WR-5	WR-5
Thickness of circular waveguide	0.374 mm (L_1)	0.3 mm (L_1)
Radius of circular waveguide	0.97 mm (R_1)	0.97 mm (R_1)
Radius of sapphire dielectric	1.97 mm (R_0)	2 mm (R_0)
Thickness of sapphire dielectric	0.2 mm (L_0)	0.2 mm (L_0)

The equivalent circuit theory is not an accurate theoretical calculation, and the designed structural parameters can only be used as an initial value. The specific size of the over-size pill-box window should be optimized by 3D electromagnetic simulation software (HFSS). In the simulation, the dielectric constant of sapphire is 9.4, and the dielectric loss tangent is 0.006. In the high-frequency pill-box window experiment process, the error capacity of the circular waveguide length (L_1) is the biggest factor affecting the pill-box window test results. The optimized condition is to ensure that the transmission bandwidth of the pill-box window is 215–225 GHz, which maximizes the tolerance of L_1. The specific optimization process is that the thickness of the sapphire and the radius of the circular waveguide remain unchanged, the radius of the different dielectric sheets are taken, and the transmission and reflection of different L_1 values ($L_0 < L_1 <$ 0.44 mm) at 220 GHz are scanned. This paper selected five different R_0 values, and recorded the transmission loss and reflection corresponding to different L_1 values at 220 GHz, under the condition that the maximum reflection in the 215–225 GHz band is less than -15 dB. The results are shown in the Figure 3. In the theoretical calculation of the traditional pill-box window, the size of the rectangular waveguide, and the radius of the circular waveguide and the dielectric sheet are determined by the operating frequency. Equation (2) is an equation about the length of the circular waveguide and the thickness of the dielectric sheet. The same theoretical calculation and simulation optimization are carried out in the traditional medicine box window. The results of the same simulation optimization of the traditional pill-box window are shown in Figure 3.

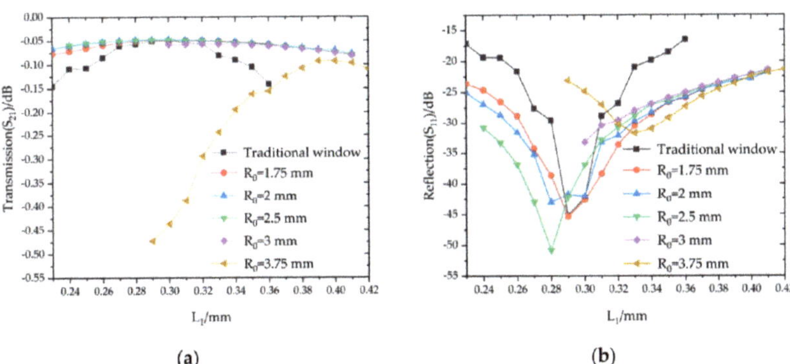

Figure 3. (a) The transmission loss and (b) reflection simulation results of pill-box windows values with different L_1.

From Figure 3a, the radii of the sapphire sheets of the over-size pill-box window are 1.75, 2, 2.5 and 3 mm, the change in the value of L_1 has less influence on the transmission loss, and it is also less than the influence of traditional pill-box windows. When the radius of the dielectric sheets of the over-size pill-box window is larger (R_0 = 3.75 mm), the change in the value of L_1 has a greater impact on the transmission loss. In Figure 3b, it can be seen that the tolerance of L_1 is different for sapphire sheets with different radii. Under the condition that the reflection of the pill-box window at 220 GHz is less than -30 dB and the maximum reflection in the 215–225 GHz frequency band is less than -15 dB, when the radius of the sapphire sheet of the over-size pill-box window is 1.75, 2, and 2.5 mm, the value change range of L_1 is 0.08 mm. The range of L_1 value of the traditional pill-box window under the same reflection condition is 0.03 mm. It can be seen that the radius of the sapphire sheet of the over-size pill-box window is set to an appropriate value, which can increase the tolerance of the structural parameters of the window system. The final radius of the sapphire in this paper is selected to be 2 mm because of the brazing and miniaturization of the pill-box window. The final structural parameters of the box window are shown in Table 1.

In a simulation template with PEC (Perfect Electric Conductor) as the background material, a pill-box window model, as shown in Figure 4, is established, and the rectangular waveguide openings at the upper and lower ends are set as the excitation source. The simulation result and theoretical calculation results of the transmission characteristics of the pill-box window are shown in Figure 5. It can be seen that the over-size pill-box window calculated by the equivalent circuit theory has a reflection of -42.5 dB at 220 GHz, but its bandwidth is relatively narrow. The optimized over-size pill-box window structure has good transmission performance in the simulation. The reflection in the 210–230 GHz frequency band is less than -30 dB, and the maximum transmission loss is -0.2 dB in the frequency band.

In order to provide the corresponding error and accuracy of the experimental assembly and processing, it is necessary to perform error analysis on the structural parameters of the over-size pill-box window. It can be concluded from Figure 3 that the structural parameters L_1 and R_0 of the over-size pill-box window have a large tolerance. The error analysis is mainly on the radius of the circular waveguide (R_1) and the thickness of the dielectric sheet (L_0). By keeping the other structural dimensions of the pill-box window unchanged, the change in the value of R_1, and the transmission loss and reflection of the pill-box window are shown in Figure 6. When the radius of the circular waveguide (R_1) changes within 0.96–0.98 mm, the reflection of the pill-box window in the frequency band 215–225 GHz is less than -20 dB, and its transmission loss changes little. The same simulation error analysis is performed on the thickness of the dielectric sheet (L_0), and the result is shown in Figure 7. When the variation range of L_0 is 0.194–0.206 mm, the reflection parameters of the pill-box window are less than -20 dB in the 215–225 GHz band, and the transmission loss is less than -0.15 dB. It can be seen from the above that the tolerance difference in the two parameters R_1 and L_0 is ±0.01 mm and ±0.006 mm.

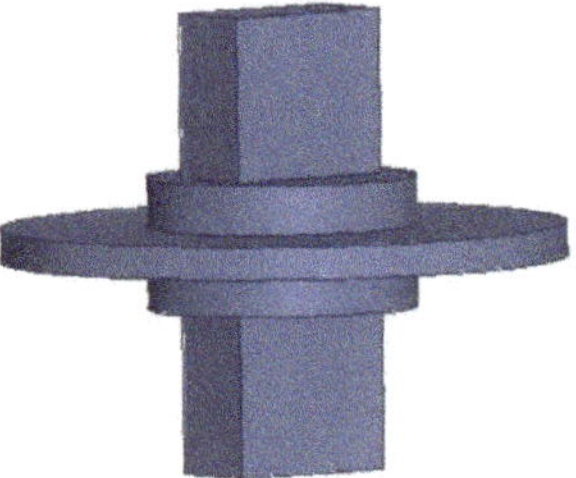

Figure 4. The simulation structure of over-size pill-box window.

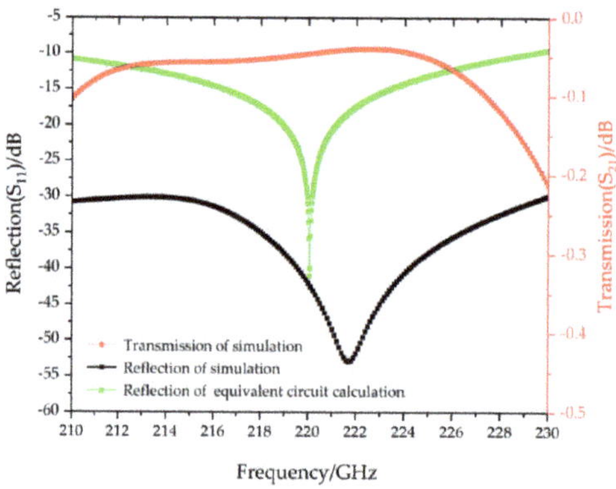

Figure 5. Theoretical and simulated transmission characteristics.

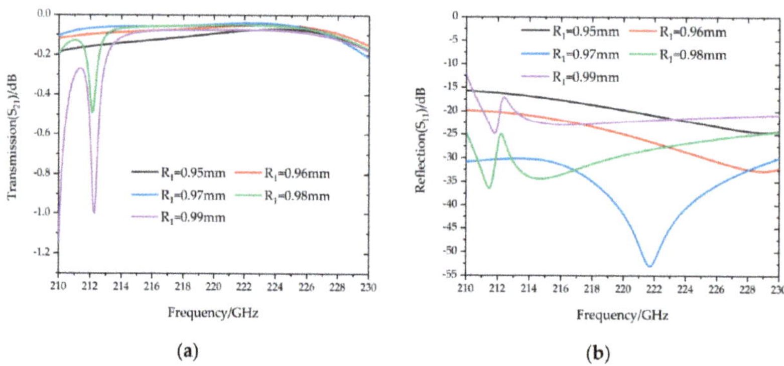

Figure 6. (**a**) The transmission loss and (**b**) reflection of pill-box windows values with different R_1.

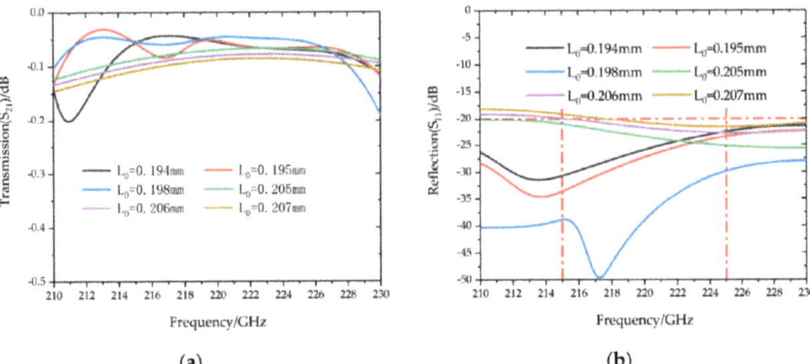

Figure 7. (**a**) The transmission loss and (**b**) reflection of pill-box windows values with different L_0.

3. Experiment Results

In the preliminary test of the over-size pill-box window, we found that the test size of the individual components was within the tolerance of the error analysis, but the test results after assembly and welding were not ideal. Because the over-size pill-box window being tested is composed of many parts, the final test result is determined by the errors of all components. In order to increase the utilization rate of over-size pill-box window components and the yield of over-size pill-box windows, this paper optimizes the testing process, as shown in Figure 8. The components of the over-size pill-box window are cleaned and dimensioned, and components with qualified dimensions are selected for assembly, and then the first test is performed on the vector network analyzer (CEYAER AV3672C (Shandong, China)). The sapphire sheet of the over-size pill-box window that passes the test is selected for metallization, and the second transmission test is performed after assembly with the original components. The over-size pill-box window that passes the second transmission test is selected to complete the final assembly and brazing. Finally, the third transmission test is performed. In this test scheme, the first two transmission tests can eliminate unqualified assembly components during the test process, minimize the waste of the pill-box window components, and ensure the qualification of the over-size pill-box window components for the final transmission test to increase the yield of over-size pill-box windows.

It is worth mentioning that, when formulating a pill-box window soldering plan, the solder should be isolated from the cavity waveguide in the pill-box window, that is, the solder slot in the soldering plan should be a closed space, as shown in Figure 9a. The deformation of the solder in the high-temperature furnace under the high pressure of the fixture should be controlled. This can reduce the influence of transmission characteristics of the over-size pill-box window by solder deformation. According to the structural parameters optimized by the simulation, the components of the over-size pill-box window are processed; the overall pill-box window after brazing is shown in Figure 9b. The experimental system of the over-size pill-box window is shown in Figure 8.

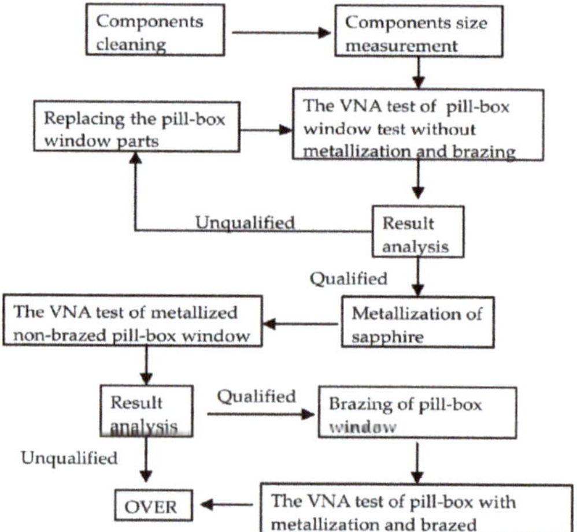

Figure 8. The test flow chart for the over-size pillbox window.

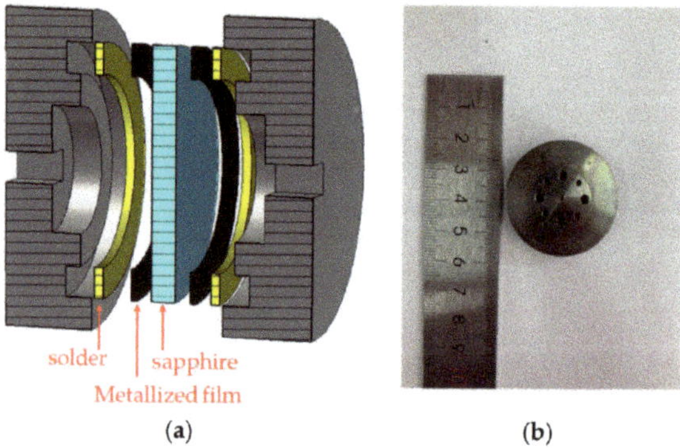

Figure 9. (a) The b razing diagram of the window system (b) The pillbox windows after brazing assembly.

From the experimental process of the pill-box window, it can be seen that each successful window needs three tests on the VNA. In the first test, the non-metalized dielectric window sheet and other components were assembled into a pill-box window system with screws and pins. The test results on VNA are shown in Figure 10. It can be seen from the figure that there is a large reflection at 214.3 GHz, the transmission loss reaches −2 dB, and the spurious modes are produced at the frequency of 221.9 GHz. The reason for this phenomenon may be that the annular air grooves on both sides of the tested pill-box window dielectric sheet affect the transmission characteristics of the window system. This can be verified by a simulation calculation. The ring-shaped hollow groove with a ring width of 0.5 mm and a ring height of 0.2 mm on both sides of the dielectric sheet are added to the simulation calculation, and the result is shown in Figure 11. There is a larger reflection point at 213.5 GHz in the simulation result, which is 0.9 GHz different from the frequency of the reflection point in the test result. The two results are quite similar. It can be determined that this reflection point is the influence of the ring-shaped groove on the transmission of the pill-box window. The test results have good overall transmission characteristics after removing the spurious mode points in the 215–225 GHz frequency band, and are highly similar to the simulation results. This shows that the processing error of the tested pill-box window parts is small, and the next test can be carried out.

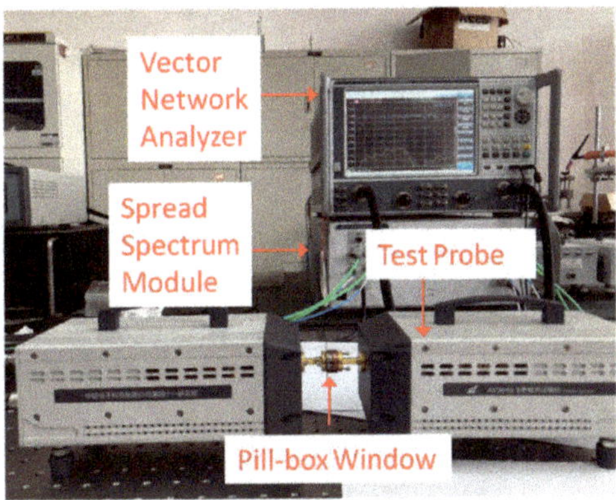

Figure 10. Experimental testing of the pill-box windows.

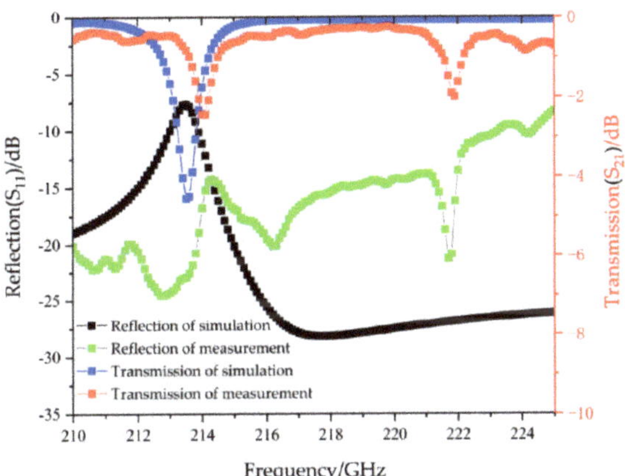

Figure 11. Results of the non-metallization and non-brazed pill-box windows.

After the first test of the pill-box window, the dielectric window sheet is metalized and then assembled and tested. In this case, the ring-shaped solder grooves on both sides of the dielectric sheet are blocked by the metalized film, and the transmission effect on the pill-box window is negligible. With the support of mature metallization and brazing technology, the results of the last two tests are basically the same. The final test results are shown in Figure 12. It can be seen from the figure that there is a large gap between the tested pill-box window transmission coefficient and the simulated value, but the numerical trend is very similar. In the design frequency band 215–225 GHz, the over-size pill-box window has good transmission characteristics, its maximum transmission loss is −1 dB, the overall transmission loss is close to −0.5 dB, and the overall reflection is less than −11 dB. Its characteristics fully meet the parameter requirements as an input window.

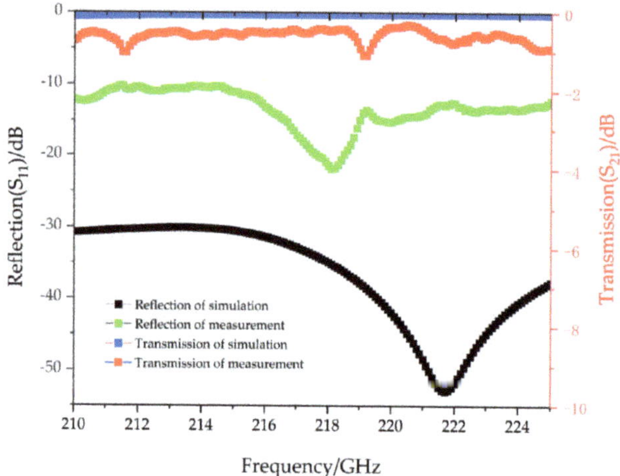

Figure 12. Results of the metallization and brazed pill-box windows.

4. Conclusions

This paper introduces the design, production and testing of a pill-box window developed for G band VEDs. The tolerance of the circular waveguide length (L_1) of the pill-box window in the different radius mediums of the improved pill-box window is discussed, which effectively provides a higher tolerance of the box-shaped window to the length of the circular waveguide. This provides a new idea for the selection of the radius of the dielectric sheet of this type window. After the theoretical calculation and simulation optimization, the reflection of the box window in the 210–230 GHz frequency band was less than −30 dB, and the maximum transmission loss was −0.2 dB. The problems of the experiment process and optimized the test process are summarized. The test results of pill-box windows without metallization and brazing are shown. The simulation analysis on test results was performed to determine the eligibility of pill-box window components. The pill-box window was within the design frequency band of 215–225 GHz, the overall transmission loss was close to −0.5 dB, and the overall reflection was −11 dB. The test results show that this over-size pill-box window is a design that can be applied to the vacuum window of a sub-THz VEDs, and has development potential at higher terahertz region frequencies.

Author Contributions: T.Y. and W.F. contributed to the overall study design, analysis, computer simulation, and writing of the manuscript. X.G., D.L., J.X., C.Z., X.Y., and Y.Y. provided technical support and revised the manuscript. All authors have read and agreed to the published version of the manuscript.

Funding: This work was supported in part by National Key Research and Development Program of China under 2019YFA0210202, in part by the National Natural Science Foundation of China under Grant 61971097 and 6201101342, and in part by the Terahertz Science and Technology Key Laboratory of Sichuan Province Foundation under Grant THZSC201801.

Acknowledgments: The authors gratefully acknowledge Yin Huang and Weirong Deng for their kind assistance on engineering design and assembling.

Conflicts of Interest: The authors declare no conflict of interest.

References

1. Lin, M.C.; Chung, H.M.; Burke, A. An analytical approach for the fast design of high-power waveguide windows. *Wave Motion* **2003**, *37*, 183–188. [CrossRef]
2. Donaldson, C.R.; He, W.; Zhang, L.; Cross, A.W. A W-band multilayer microwave window for pulsed operation of gyro-devices. *IEEE Microw. Wirel. Compon. Lett.* **2013**, *23*, 237–239. [CrossRef]
3. Cook, A.M.; Joye, C.D.; Kimura, T.; Wright, E.L.; Calame, J.P. Broadband 220-GHz vacuum window for a traveling-wave tube amplifier. *IEEE Trans. Electron. Devices* **2013**, *60*, 1257–1259. [CrossRef]
4. Ao, H.; Asano, H.; Naito, F.; Ouchi, N.; Tamura, J.; Takata, K. Impedance matching of pillbox-type RF windows and direct measurement of the ceramic relative dielectric constant. *Nucl. Instrum. Methods Phys. Res. Sect. A Accel. Spectrom. Detect. Assoc. Equip.* **2014**, *737*, 65–70. [CrossRef]
5. Zhang, L.; Donaldson, C.R.; Cross, A.W.; He, W. A Pillbox window with impedance Matching Sections for a W-Band Gyro-TWA. *IEEE Electron Device Lett.* **2018**, *39*, 1081–1084. [CrossRef]
6. Yang, T.; Guan, X.; Fu, W.; Lu, D.; Zhang, C.; Xie, J.; Yuan, X.; Yan, Y. Investigation on Symmetric and Asymmetric Broadband Low-Loss W-Band Pillbox Windows. *Electronics* **2020**, *9*, 2060. [CrossRef]
7. Cai, J.; Hu, L.; Ma, G.; Chen, H.; Jin, X.; Chen, H. Theoretical and Experimental Study of the Modified Pill-Box Window for the 220-GHz Folded Waveguide BWO. *IEEE Trans. Plasma Sci.* **2014**, *42*, 3349–3357. [CrossRef]
8. Arai, H.; Goto, N.; Ikeda, Y.; Imai, T. An analysis of a vacuum window for lower hybrid heating. *IEEE Trans. Plasma Sci.* **1986**, *14*, 947–954. [CrossRef]
9. Liu, S. A RF window for broadband millimeter wave tubes. *Int. J. Infrared Millim. Waves* **1996**, *17*, 121–126. [CrossRef]

 *electronics*

Article

Investigation on Symmetric and Asymmetric Broadband Low-Loss W-Band Pillbox Windows

Tongbin Yang [1], Xiaotong Guan [2,3], Wenjie Fu [1,3,*], Dun Lu [1], Chaoyang Zhang [1], Jie Xie [1], Xuesong Yuan [1,3] and Yang Yan [1,3]

[1] School of Electronic Science and Engineering, University of Electronic Science and Technology of China, Chengdu 610054, China; yangtongbin@std.uestc.edu.cn (T.Y.); ludun@std.uestc.edu.cn (D.L.); chaoyangzhang@std.uestc.edu.cn (C.Z.); xiejie@std.uestc.edu.cn (J.X.); yuanxs@uestc.edu.cn (X.Y.); yanyang@uestc.edu.cn (Y.Y.)

[2] School of Physics, University of Electronic Science and Technology of China, Chengdu 610054, China; guanxt@uestc.edu.cn

[3] Terahertz Science and Technology Key Laboratory of Sichuan Province, University of Electronic Science and Technology of China, Chengdu 610054, China

* Correspondence: fuwenjie@uestc.edu.cn

Received: 5 November 2020; Accepted: 1 December 2020; Published: 3 December 2020

Abstract: In order to develop wide-band low-loss windows for W-band vacuum electronic devices and easily fabricate them, symmetric and asymmetric pillbox windows are investigated and reported in this paper. A symmetric pillbox window and an asymmetric pillow-box window were designed, simulation optimized, fabricated, and tested. The initial parameters for the two pillbox windows were designed by equivalent circuit theory. Computer simulation technology (CST) three-dimensional (3D) electromagnetic simulation software was used to verify and optimize the design. Because of the uncontrollability of welding during the experiment, this article provides two solutions. One is to measure and reprocess the symmetrical pillbox window with the dielectric sheet welded to reduce the influence of welding on the measurement results; the other is an asymmetrical box window which is designed to avoid the error caused by the welding of the box window. The best experimental results for the symmetric pillbox window were $|S_{21}|$ close to 1 dB and reflection parameter $|S_{11}|$ close to 10 dB in the frequency range of 77–110 GHz. The experimental results for the asymmetric pillbox window were $|S_{21}| < 1$ dB nearly in the frequency range of 76–109.5 GHz. The experimental results show that both solutions efficiently complete the design of broadband pillbox windows and would potentially be operated in the gigahertz millimeter-wave region.

Keywords: pillbox window; wide-band; W-band; low loss

---

1. Introduction

Microwave vacuum tubes, such as traveling wave tubes (TWTs), klystrons, magnetrons, and gyrotrons, are important electronic devices. As high-power microwave radiation sources, vacuum tubes are widely used in microwave applications, including wireless communication, high-resolution radar and imaging [1], and industrial heating. In the research of microwave vacuum tubes, the window system is used to isolate the vacuum and atmosphere, which is an indispensable component. Diverse window configurations have been reported, such as the single-disc [2], multi-disc [3], and pillbox [4–7] configurations. Compared with other types of window systems, the pillbox window system has the advantages of being wideband and easy to braze, having large power capacity, and so on. It has been widely used in window design [8–10].

In the research of W-band microwave vacuum windows, great achievements have been made. Table 1 lists the results of microwave windows that have been experimentally tested in recent years.

It can be seen from Table 1 that the pillbox window can be applied to various frequency bands of microwaves and millimeter waves. In W-band, the working frequency bands in most of the measured results for pillbox windows were concentrated within 90–100 GHz [6]. Bandwidth is one of the most important indicators of a microwave window and directly affects the performance of vacuum devices. This paper extends the working band of pillbox windows to 76–110 GHz, which is suitable for microwave devices in the full W-band.

Table 1. Some achievements in vacuum microwave windows.

Institution	Design Bandwidth	Measured Bandwidth	Window Type
University of Strathclyde [6]	10 GHz (90–100 GHz)	10 GHz (90–100 GHz)	Pillbox window
Seoul National University [8]	9 GHz (93–102 GHz)	8 GHz (93.2–101.2 GHz)	Pillbox window
Vacuum Electronics National Laboratory [11]	20 GHz (82–102 GHz)	10 GHz (92–102 GHz)	Rectangular waveguide window
China Academy of Engineering Physics [12]	20 GHz (130–150 GHz)	13 GHz (132–145 GHz)	Pillbox window
Tsinghua University [13]	10 GHz (215–225 GHz)	10 GHz (215–225 GHz)	Pillbox window

The traditional pillbox window system consists of three distinct parts: two symmetrical rectangular waveguides, a straight cylindrical waveguide, and a cylindrical waveguide filled with dielectric. The dielectric window is brazed in a straight circular waveguide. At low working frequency, this type of pillbox window has many advantages and is applied in many products. However, when the operation frequency is up to the sub-millimeter-wave and millimeter-wave region, the yield and reliability of this pillbox window are very low in fabrication and assembly. To improve the success rate of the pillbox window, the traditional window is tested and analyzed in this paper. The reasons for low assembly efficiency are as follows: (1) With increasing operation frequency, the structural parameters of the pillbox window decrease. In the sub-millimeter-wave and millimeter-wave region, it is hard to ensure that the dielectric window is centered and vertical in the straight circular waveguide; thus, it is easy to simulate spurious modes, reducing the bandwidth. (2) Symmetrical pillbox windows cannot be tested before they are brazed. If the test results are not satisfactory, the window is scrapped. To solve these problems, this paper provides two solutions. One is to measure and process the symmetrical pillbox window with the dielectric sheet welded to avoid the influence of welding on the measurement results; the other is an asymmetric pillbox window that was investigated (these results have been published in another paper [14]), the structure of which is shown in Figure 1b. In comparison to the conventional symmetric pillow-box window which is shown in Figure 1a, we see that the diameter of the two sections of the cylindrical waveguide is different in the asymmetric pillow-box window. This system can nicely restrain the variations caused by the brazing process and ensure consistency between the experimental results and design. It can also be roughly tested before brazing to correct the structural parameters.

This paper is organized as follows. Section 2 introduces the selection of the window dielectric material, equivalent circuit theory of the symmetrical pillbox window, simulation results, and structural parameters of the two kinds of pillbox window. Section 3 is devoted to experiments, analysis of experiment results, and improvement of the symmetrical pillbox window and the experimental result of the asymmetrical pillbox window. Section 4 concludes the paper.

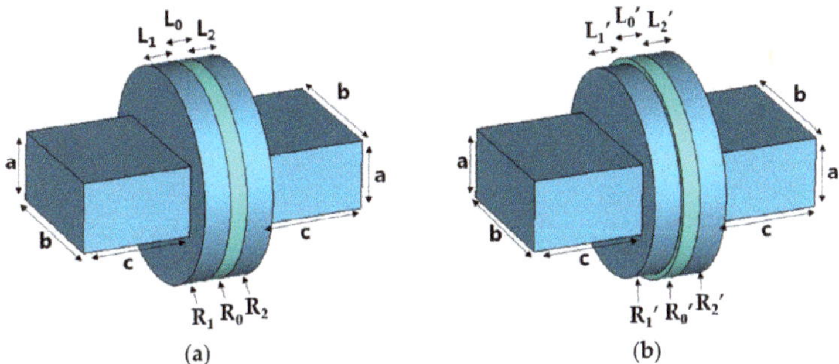

Figure 1. Structure of the (a) symmetrical pillbox window and (b) asymmetrical pillbox window.

2. Model Design and Simulation

In the design of the pillbox window, the first step is the determination of the material of the window piece. Sapphire was finally selected in this investigation; its reference factors include high strength and small electric loss tangent ($<1 \times 10^{-4}$), and the metalized technology is mature. Strictly speaking, the relative permittivity values in the horizontal and vertical directions of sapphire are different. Under the condition of TE (Transverse electric) mode transmission, the difference in the dielectric constant of sapphire has little effect on the transmission of the pillbox window [13]. The relative permittivity of the transmission is 9.2, according to material reports.

There are some calculation methods used in the design of symmetrical traditional box windows, including the impedance matching approach [6], the method of moments [13], and the equivalent circuit method [7,13,15,16]. In this paper, the equivalent circuit method was used. The transmission modes in the pillbox window are the TE_{10} mode of the rectangular waveguide and the TE_{11} mode of the cylindrical waveguide. An equivalent circuit diagram of the pillbox window system is shown in Figure 2. The transfer matrix of the window system is divided into nine parts, under the condition of single-mode transmission [13]. In the following theoretical calculation part, we discuss the calculation process of the symmetric window.

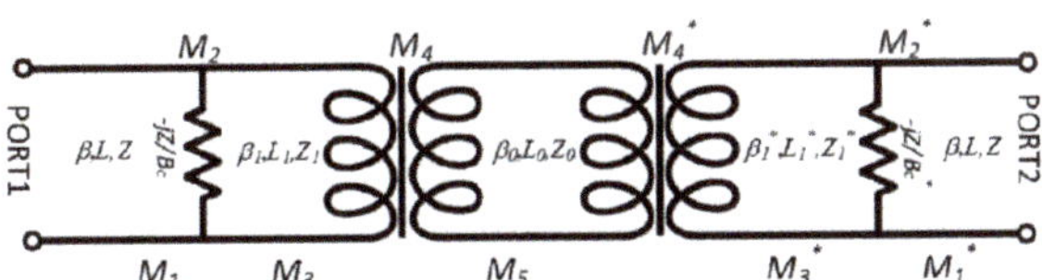

Figure 2. Diagram of the equivalent circuit.

In Figure 2, L is the length of the rectangular waveguide; Bc is the equivalent susceptance of the transition from the rectangular waveguide to the circular waveguide; L_0, L_1, and L_1^* are the length of the cylindrical waveguide; β is the propagation constant of the TE_{10} mode in the rectangular waveguide; Z is the characteristic impedance of the rectangular waveguide; β_1 is the propagation constant of the TE_{11} mode in the cylindrical waveguide; Z_1 and Z_1^* are the characteristic impedance of the cylindrical waveguide; β_0 is the propagation constant of the TE_{11} mode in a dielectric circular waveguide; Z_0 is the characteristic impedance of the dielectric waveguide; and M is the matrix of each waveguide and the connection part.

The transfer matrix of the window system can be obtained by multiplying the nine parts of the transfer matrix, shown as

$$M = M_1 \bullet M_2 \bullet M_3 \bullet M_4 \bullet M_5 \bullet M_4^* \bullet M_3^* \bullet M_2^* \bullet M_1 \tag{1}$$

Substituting the transfer matrix of each part into Equation (1), the transfer matrix can be written as [16–19]

$$
M = \begin{bmatrix} \cos(\beta L) & -jZ\sin(\beta L) \\ -\frac{j}{Z}\sin(\beta L) & \cos(\beta L) \end{bmatrix} \bullet \begin{bmatrix} 1 & 0 \\ -\frac{jB_c}{Z} & 1 \end{bmatrix} \bullet \begin{bmatrix} \cos(\beta_1 L_1) & -jZ_1\sin(\beta_1 L_1) \\ -\frac{j}{Z_1}\sin(\beta_1 L_1) & \cos(\beta_1 L_1) \end{bmatrix} \bullet \begin{bmatrix} R_0/R_1 & 0 \\ 0 & R_1/R_0 \end{bmatrix} \bullet \begin{bmatrix} \cos(\beta_0 L_0) & -jZ_0\sin(\beta_0 L_0) \\ -\frac{j}{Z_0}\sin(\beta_0 L_0) & \cos(\beta_0 L_0) \end{bmatrix}
$$
$$
\bullet \begin{bmatrix} R_2/R_0 & 0 \\ 0 & R_0/R_2 \end{bmatrix} \bullet \begin{bmatrix} \cos(\beta_1 L_1) & -jZ_1\sin(\beta_1 L_1) \\ -\frac{j}{Z_1}\sin(\beta_1 L_1) & \cos(\beta_1 L_1) \end{bmatrix} \bullet \begin{bmatrix} 1 & 0 \\ -\frac{jB_c}{Z} & 1 \end{bmatrix} \bullet \begin{bmatrix} \cos(\beta L) & -jZ\sin(\beta L) \\ -\frac{j}{Z}\sin(\beta L) & \cos(\beta L) \end{bmatrix} \tag{2}
$$

where Bc can be written as [18,19]

$$
\begin{cases}
B_c = \frac{\beta b}{2\pi}\left\{ 2\ln\frac{R_1{}^2-b^2}{4R_1 b} + \left(\frac{b}{R_1}+\frac{R_1}{b}\right) + 2\sum_{n=1}^{\infty}\frac{\sin^2 n\varphi}{n^3\varphi^2}\delta_{2n} \right\} \\
\delta_{2n} = \frac{1}{\sqrt{1-\left(\frac{\beta R_1}{2n\pi}\right)^2}} - 1,\ \varphi = \frac{\pi b}{R_1}
\end{cases} \tag{3}
$$

where b is the short side length of a rectangular waveguide, and $Bc = Bc^*$.

In the window system, the length of the rectangular waveguides at both sides only influences the phase of the transfer matrix, L is set to zero, and $R_0 = R_1 = R_2$. The transfer matrix of the symmetric window system can be written as

$$
\begin{aligned}
M &= M_2 \bullet M_3 \bullet M_5 \bullet M_3^* \bullet M_2^* \\
&= \begin{bmatrix} 1 & 0 \\ -\frac{jB_c}{Z} & 1 \end{bmatrix} \bullet \begin{bmatrix} \cos(\beta_1 L_1) & -jZ_1\sin(\beta_1 L_1) \\ -\frac{j}{Z_1}\sin(\beta_1 L_1) & \cos(\beta_1 L_1) \end{bmatrix} \bullet \begin{bmatrix} \cos(\beta_0 L_0) & -jZ_0\sin(\beta_0 L_0) \\ -\frac{j}{Z_0}\sin(\beta_0 L_0) & \cos(\beta_0 L_0) \end{bmatrix} \\
&\bullet \begin{bmatrix} \cos(\beta_1 L_1) & -jZ_1\sin(\beta_1 L_1) \\ -\frac{j}{Z_1}\sin(\beta_1 L_1) & \cos(\beta_1 L_1) \end{bmatrix} \bullet \begin{bmatrix} 1 & 0 \\ -\frac{jB_c}{Z} & 1 \end{bmatrix}
\end{aligned} \tag{4}
$$

According to the relation between the transmission matrix and scattering matrix, the scattering matrix S could be written as

$$
S = \begin{bmatrix} \frac{M_{22}+M_{12}-M_{21}-M_{11}}{M_{11}+M_{12}+M_{21}+M_{22}} & \frac{2}{M_{11}+M_{12}+M_{21}+M_{22}} \\ 2\frac{M_{11}M_{22}-M_{12}M_{21}}{M_{11}+M_{12}+M_{21}+M_{22}} & \frac{M_{11}+M_{12}-M_{21}-M_{22}}{M_{11}+M_{12}+M_{21}+M_{22}} \end{bmatrix} \tag{5}
$$

The S-parameter ideal design goal of the symmetric window is $S_{12} = S_{21}$, $S_{11} = S_{22} = 0$; thus, the structural parameters satisfy the equations

$$
\begin{cases}
\tan\beta_1 L_1 = A/B \pm \sqrt{(A/B)^2 - C/B} \\
(A/B)^2 - C/B \geq 0 \\
A = \left(-\frac{Z}{Z_1} + \frac{Z_1}{Z}B_c^2 + \frac{Z}{Z_1}\right)\cos\beta_0 L_0 + \left(\frac{Z_1}{Z_0}B_c^2 + \frac{Z_0}{Z_1}B_c\right)\sin\beta_0 L_0 \\
B = -2B_c\cos\beta_0 L_0 + \left(\frac{Z_1^2}{ZZ_0}B_c^2 - \frac{ZZ_0}{Z_1^2} + \frac{Z_1^2}{ZZ_0}\right)\sin\beta_0 L_0 \\
C = 2B_c\cos\beta_0 L_0 + \left(\frac{Z}{Z_0} - \frac{Z_0}{Z}B_c^2 - \frac{Z_0}{Z}\right)\sin\beta_0 L_0
\end{cases} \tag{6}
$$

According to the frequency of the pillbox window, the WR-10 (Waveguide name of EIA standard) standard rectangular window was selected. The three structural parameters of the symmetrical pillbox window, L_0, L_1, and R_0, need to satisfy Equation (6). When the center frequency of the design band is 92.5 GHz, the three parameters have infinitely many solutions that satisfy the equation. To obtain the ideal parameter structure, the values of the three structural parameters need to be limited. Considering the sealing and mechanical strength of the window, the dielectric sheet needs to have a larger thickness, and considering the matching of the window system, the window sheet needs to be thinner. Based on the above reasoning, L_0 is between one-half of the waveguide wavelength (0.55 mm) and one-quarter of the waveguide wavelength (0.275 mm) (0.275 mm $\leq L_0 \leq$ 0.55 mm). The diameter of the circular waveguide is too large, the high-order mode cannot be cut off, and the diameter of the window is too small, which will narrow the matching bandwidth. The diameter of the general box-shaped window is taken from the diagonal of the rectangular waveguide $D = \sqrt{a^2 + b^2}$. Based on the bandwidth requirements of the pillbox window in this paper, the diameter of the window can be slightly larger than D, and the scanning range can be set as $L_0 \leq L_1 \leq 1$ mm. The value ranges of the three parameters were combined to find the solution set that satisfies Equation (6). The transmission and reflections of each set of structural parameter values can be obtained by Equation (5); the parameter group with a larger bandwidth and reflection less than −10 dB was selected as shown in Table 2.

Table 2. Structural parameters of the pillbox windows.

	Equivalent Circuit	Simulation—Symmetric	Simulation—Asymmetric
Rectangular waveguide	WR-10	WR-10	WR-10
Thickness of circular waveguide	0.49 mm (L_1) 0.49 mm (L_2)	0.48 mm (L_1) 0.48 mm (L_2)	0.48 mm ($L_1{}'$) 0.48 mm ($L_2{}'$)
Radius of circular waveguide	1.9 mm (R_1) 1.9 mm (R_2)	1.8 mm (R_1) 1.8 mm (R_2)	1.8 mm ($R_1{}'$) 1.75 mm ($R_2{}'$)
Radius of sapphire dielectric	1.9 mm (R_0)	1.8 mm (R_0)	1.8 mm ($R_0{}'$)
Thickness of sapphire dielectric	0.31 mm (L_0)	0.34 mm (L_0)	0.33 mm ($L_0{}'$)

The reflection value by the equivalent circuit theory and the transmission characteristics in CST are shown in Figure 3. Except for the frequency deviation of 4 GHz, the two reflection curves have good consistency. This shows that the equivalent circuit theory is reasonable as a preliminary design tool. The equivalent circuit theory is not a very accurate mathematical model, and further optimization should rely on CST three-dimensional (3D) electromagnetic simulation software. The structural parameters were simulated and optimized with the goal of wider transmission bandwidth and smaller transmission loss. The structural parameters of the asymmetric pillbox window were optimized based on the parameters of the symmetrical-type window. Considering the machining accuracy and assembly difficulty, the diameter difference of the two circular waveguides was taken as 0.1 mm. The structural parameters are shown in Table 2.

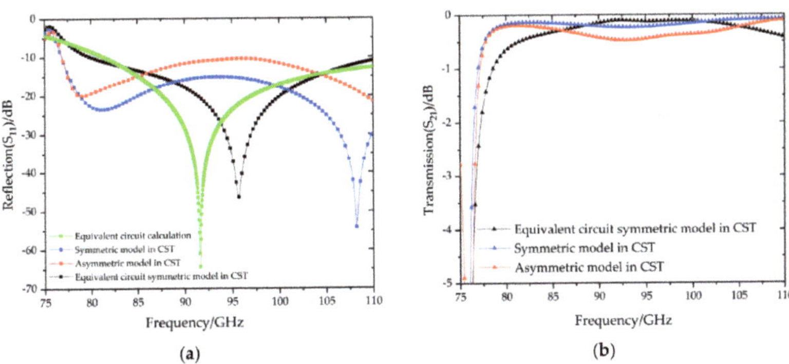

Figure 3. The optimized (**a**) S_{11} values of pillbox windows and (**b**) S_{21} values of pillbox windows.

The simulation-optimized structural model is shown in Figure 4. For input mode TE_{10}, the transmission characteristics are shown in Figure 3. In the frequency range of 78–110 GHz, the S_{21} values of the two designs were greater than −0.5 dB; the maximum transmission loss of the symmetric pillbox window was −0.23 dB, while that of the asymmetric type was −0.47 dB.

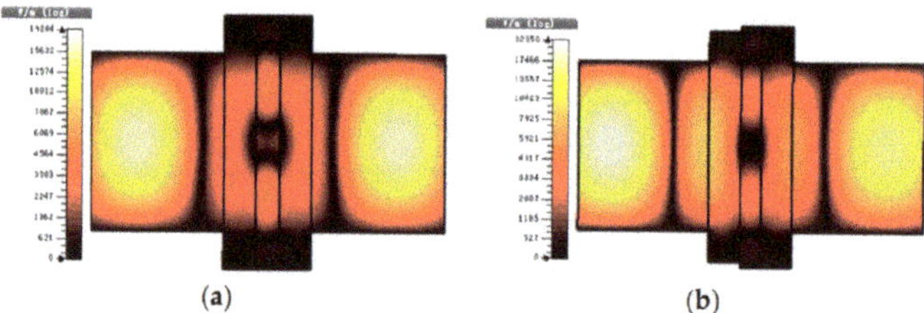

(a)   (b)

Figure 4. Simulation model of the (**a**) symmetrical pillbox window and (**b**) asymmetrical pillbox window.

3. Experiment Results

According to the optimized structural parameters (Table 2), two box-type windows were processed and assembled, as shown in Figure 5. The testing process in the vector network analyzer (CEYAER AV3672C) is shown in Figure 6.

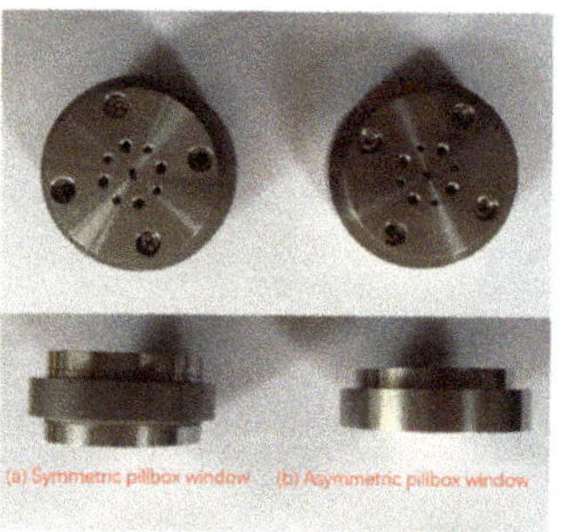

Figure 5. The pillbox windows after brazing assembly: (**a**) symmetric pillbox window and (**b**) asymmetric pillbox window.

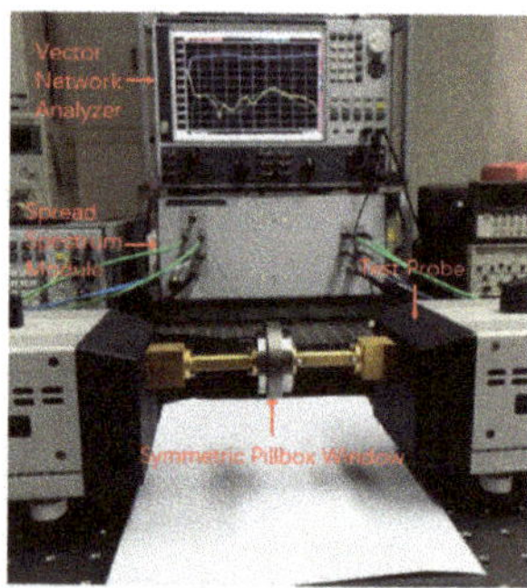

Figure 6. Experimental testing of the pillbox windows.

The experimental results of the symmetrical box window in the early stage were not ideal. The error analysis method in the literature [6] was used to analyze the experimental results. Based on the analysis results, the experimental program was revised. Because the value of S_{11} was small and its influence on the change of the structural parameters of the window system was not obvious enough, the more sensitive S_{21} was selected as the optimization target, and the random error was the structural parameter of the window. Some of the analysis results are shown in Figure 7, and the calculated tolerance of the window is shown in Table 3.

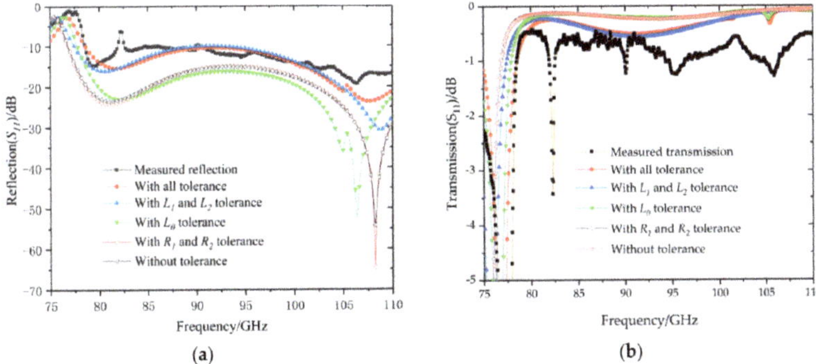

Figure 7. Measurement results and simulation analysis of the symmetric window: (**a**) S_{11} values of the pillbox window; (**b**) S_{21} values of the pillbox window.

Table 3. The calculation tolerance of the pillbox window.

Variable Name	L_0	L_1	L_2	R_0	R_1	R_2
Figure 7	−0.013 mm [1]	0.06 mm	−0.08 mm	0 mm	0.02 mm	0.02 mm
Figure 9	0.02 mm	−0.02 mm	−0.05 mm	−0.04 mm	0.01 mm	0.01 mm

[1] A negative sign indicates that the experimental size estimated by the simulation was smaller than the design size.

As shown in Figure 7, the errors of R_1, R_2, and L_0 made lesser contributions to the deviation of the measurement result and the design value, which is mainly caused by the tolerance of L_1 and L_2. The main reason is that the position of the dielectric sheet shifted during the welding process. This shift is random due to the purely manual operation of welding fixture assembly and solder filling. The test results of multiple rounds of modification of the welding fixture and assembly scheme showed that this randomness is inevitable. To reduce the influence of welding on the test process, the test plan was modified. After the windows were welded to the circular waveguide, the circular waveguides on both sides were measured and reprocessed to ensure that the values of L_1 and L_2 were within the tolerance range. The test process is shown in Figure 8. This method can effectively reduce the influence of welding on L_1 and L_2. The final test result is shown in Figure 9. By error simulation analysis of the test results, the calculation tolerance of each structural parameter was obtained, shown in Table 3. The transmission parameter $|S_{21}|$ was close to 1 dB and the reflection parameter $|S_{11}|$ was close to 10 dB in the frequency range of 77–110 GHz.

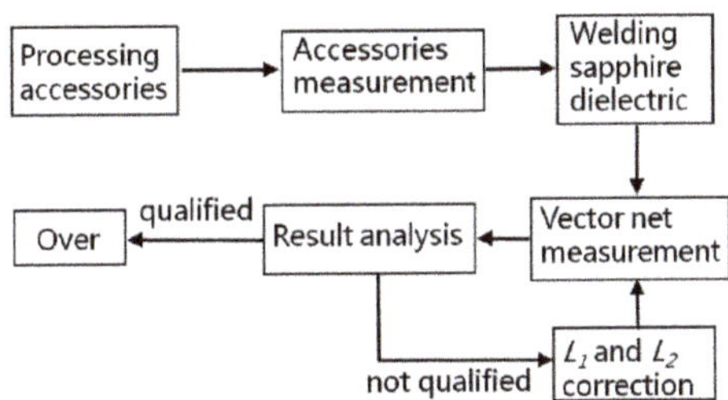

Figure 8. The test flow chart for the symmetric pillbox window.

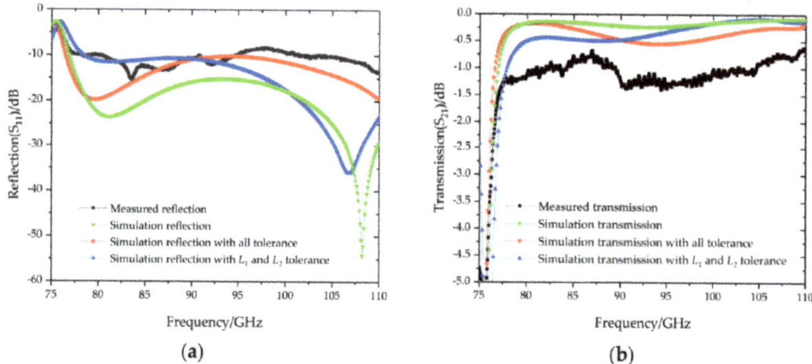

Figure 9. The measurement results of the symmetric pillbox window. (a) S_{11} values of the pillbox window; (b) S_{21} values of the pillbox window.

The asymmetric pillbox window can avoid the problems of the symmetric pillbox window in the testing. In asymmetric pillbox window testing, the window can be roughly tested before brazing to reduce the error of structural parameters. The test process of the asymmetric window is shown in Figure 10. A comparison between the two measured results and the simulation is shown in Figure 11. The frequency band of the unwelded pill window with a transmission parameter close to 0.5 dB covers 76–109.5 GHz, except for two frequency points at 87.99 GHz and 103.15 GHz. At the frequency of 87.99 GHz, S_{21} is decreased when S_{11} is increased due to reflection. At the frequency of 103.15 GHz, both S_{11} and S_{21} are decreased, which may be caused by spurious modes. Compared to the test results of the unwelded pill window, the measurement results add a spurious mode point at 107.4 GHz. The two test results have a similar frequency offset—about 2 GHz—compared to the calculation results.

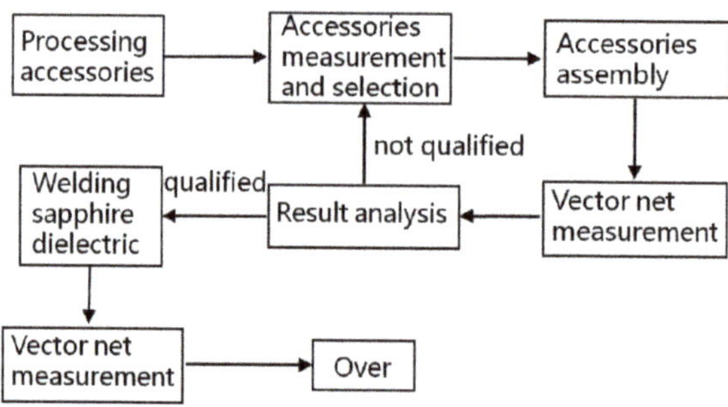

Figure 10. The test flow chart for the asymmetric pillbox window.

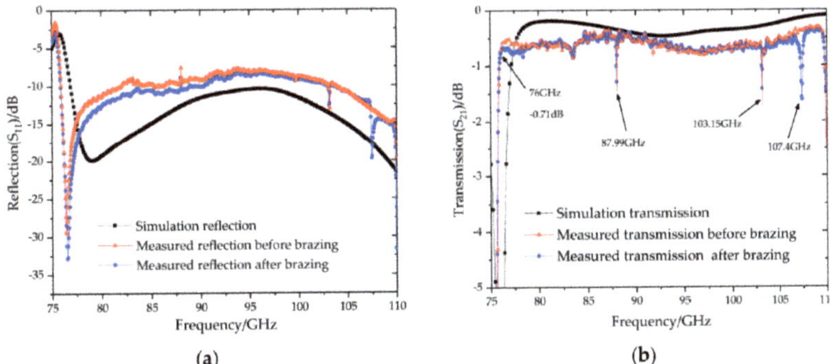

Figure 11. The measurement results of the asymmetric pillbox window: (**a**) S_{11} values of the pillbox window; (**b**) S_{21} values of the pillbox window.

4. Summary

In this paper, we report the design, manufacture, and testing of two kinds of wide-band low-loss pillbox windows for W-band vacuum electronic devices. After the experiments, analysis, and improvement, the final results for the symmetrical window were transmission parameter $|S_{21}|$ close to 1 dB and reflection parameter $|S_{11}|$ close to 10 dB in the frequency range of 77–110 GHz. Based on the experimental testing of the symmetrical-type window and the analysis of the experimental results, the asymmetrical pillbox window can reduce the errors caused by the welding process and the assembly process by using an asymmetric circular waveguide to fix the sapphire medium. The test results showed that the transmission parameter $|S_{21}|$ was <1 dB in the frequency range of 76–109.5 GHz, covering the W-band. Our test data and simulation results show that systematic tuning of the asymmetric window after fabrication can achieve broadband transmission close to the ideal design.

Both design methods can reduce the errors caused by welding and experimental processing, and both have successfully produced low-loss, wide-band box windows. The experimental methods in these two testing processes can be applied to the production of box-shaped windows in the terahertz band.

Author Contributions: T.Y. and W.F. contributed to the overall study design, analysis, computer simulation, and writing of the manuscript; X.G., D.L., J.X., C.Z., X.Y. and Y.Y. provided technical support and revised the manuscript. All authors have read and agreed to the published version of the manuscript.

Funding: This work was supported in part by the National Natural Science Foundation of China under Grant 61971097, in part by the National Key Research and Development Program of China under 2019YFA0210202, in part by the Sichuan Science and Technology Program under Grant 2018HH0136, and in part by the Terahertz Science and Technology Key Laboratory of Sichuan Province Foundation under Grant THZSC201801.

Acknowledgments: The authors gratefully acknowledge Yin Huang and Weirong Deng for their kind assistance on engineering design and assembling.

References

1. Gallerano, G.P.; Biedron, S. Overview of terahertz radiation sources. In Proceedings of the 2004 FEL Conference, Trieste, Italy, 29 August–3 September 2004; pp. 216–221.
2. Lin, M.C.; Chung, H.M.; Burke, A. An analytical approach for the fast design of high-power waveguide windows. *Wave Motion* **2003**, *37*, 183–188. [CrossRef]
3. Donaldson, C.R.; He, W.; Zhang, L.; Cross, A.W. A W-band multilayer microwave window for pulsed operation of gyro-devices. *IEEE Microw. Wireless Compon. Lett.* **2013**, *23*, 237–239. [CrossRef]

4. Cook, A.M.; Joye, C.D.; Kimura, T.; Wright, E.L.; Calame, J.P. Broadband 220-GHz vacuum window for a traveling-wave tube amplifier. *IEEE Trans. Electron Devices* **2013**, *60*, 1257–1259. [CrossRef]
5. Ao, H.; Asano, H.; Naito, F.; Ouchi, N.; Tamura, J.; Takata, K. Impedance matching of pillbox-type RF windows and direct measurement of the ceramic relative dielectric constant. *Nucl. Instrum. Methods Phys. Res. Sect. A Accel. Spectrom. Detect. Assoc. Equip.* **2014**, *737*, 65–70. [CrossRef]
6. Zhang, L.; Donaldson, C.R.; Cross, A.W.; He, W. A Pillbox window with impedance Matching Sections for a W-Band Gyro-TWA. *IEEE Electron Device Lett.* **2018**, *39*. [CrossRef]
7. Nakatsuka, H.; Yoshida, N.; Fukai, I. Three-dimensional analysis of a vacuum window connected to the waveguide. *IEEE Trans. Plasma Sci.* **1988**, *PS-16*, 416–422. [CrossRef]
8. Srivastava, A.; Kwon, O.-J.; Sattorov, M.; Sharma, A.; Tanwar, A.; Park, G.-S. Broadband THz vacuum window using impedance matching approach. In Proceedings of the 35th International Conference on Infrared, Millimeter, and Terahertz Waves, Rome, Italy, 5–10 September 2010; p. 978. [CrossRef]
9. Hill, M.E.; Callin, R.S.; Whittum, D.H. High-power vacuum window in WR10. *IEEE Trans. Microw. Theory Tech.* **2001**, *49*, 994–995. [CrossRef]
10. Liu, S. Optimum design for RF windows in millimeter-wave high power tubes. *Int. J. Infrared Millim. Waves* **2003**, *24*, 629–630. [CrossRef]
11. Hu, Y.; Feng, J.; Cai, J.; Wu, X.; Ma, S.; Qu, B.; Zhang, J.; Chen, T. A Broadband Microwave Window for W-band TWT. In Proceedings of the 2008 IEEE International Vacuum Electronics Conference, Monterey, CA, USA, 22–24 April 2008. [CrossRef]
12. Wang, Y.; Chen, Z. Investigation of 0.14THz Pill-box Window for Folded Waveguide TWT. In Proceedings of the International Vacuum Electronics Conference (IVEC), Paris, France, 21–23 May 2013. [CrossRef]
13. Cai, J.; Hu, L.; Ma, G.; Chen, H.; Jin, X.; Chen, H. Theoretical and Experimental Study of the Modified Pill-Box Window for the 220-GHz Folded Waveguide BWO. *IEEE Trans. Plasma Sci.* **2014**, *20*. [CrossRef]
14. Yang, T.; Lu, D.; Fu, W.; Yan, Y.; Guan, X. A broadband low-loss W-band Pill-box Window. In Proceedings of the 2019 International Vacuum Electronics Conference (IVEC), Busan, Korea, 28 April–1 May 2019. [CrossRef]
15. Arai, H.; Goto, N.; Ikeda, Y.; Imai, T. An analysis of a vacuum window for lower hybrid heating. *IEEE Trans. Plasma Sci.* **1986**, *PS-14*, 947–954. [CrossRef]
16. Liu, S. A RF window for broadband millimeter-wave tubes. *Int. J. Infrared Millim. Waves* **1996**, *17*, 121–126. [CrossRef]
17. Marcuvitz, N. *Waveguide Handbook*; Peregrinus: Stevenage, UK, 1986.
18. Lewin, L. *Advanced Theory of Waveguides*; Iliffe & Sons: London, UK, 1951.
19. Wade, J.D.; Macphie, R.H. Scattering at circular-to-rectangular waveguide junctions. *IEEE Trans. Microw. Theory Tech.* **1986**, *34*, 1085–1091. [CrossRef]

Publisher's Note: MDPI stays neutral with regard to jurisdictional claims in published maps and institutional affiliations.

 *electronics*

Article

Ultra-High Velocity Ratio in Magnetron Injection Guns for Low-Voltage Compact Gyrotrons

Dun Lu [1], Wenjie Fu [1,*], Xiaotong Guan [2], Tongbin Yang [1], Chaoyang Zhang [1], Chi Chen [1], Meng Han [1] and Yang Yan [1]

[1] Terahertz Research Center, University of Electronic Science and Technology of China, Chengdu 610054, China; ludun@std.uestc.edu.cn (D.L.); yangtongbin@std.uestc.edu.cn (T.Y.); zhangchaoyang@std.uestc.edu.cn (C.Z.); chenchi03@163.com (C.C.); 201711040132@std.uestc.edu.cn (M.H.); yanyang@uestc.edu.cn (Y.Y.)

[2] School of Physics, University of Electronic Science and Technology of China, Chengdu 610054, China; guanxt@uestc.edu.cn

* Correspondence: fuwenjie@uestc.edu.cn

Received: 5 September 2020; Accepted: 24 September 2020; Published: 28 September 2020

Abstract: Low-voltage compact gyrotron is under development at the University of Electronic Science and Technology of China (UESTC) for industrial applications. Due to the low operating voltage, the relativistic factor is weak, and interaction efficiency could not be high. Therefore, a magnetron-injection gun (MIG) with an extremely high-velocity ratio α (around 2.5) is selected to improve the interaction efficiency. As beam voltage drops, space charge effects become more and more obvious, thus a more detailed analysis of velocity-ratio α is significant to perform low-voltage gyrotrons, including beam voltage, beam current, modulating voltage, depression voltage, cathode magnetic field, and magnetic depression ratio. Theoretical analysis and simulation optimization are adopted to demonstrate the feasibility of an ultra-high velocity ratio, which considers the space charge effects. Based on theoretical analysis, an electron gun with a transverse to longitudinal velocity ratio 2.55 and velocity spread 9.3% is designed through simulation optimization. The working voltage and current are 10 kV and 0.46 A with cathode emission density 1 A/cm^2 for a 75 GHz hundreds of watts' output power gyrotron.

Keywords: velocity ratio; velocity spread; low-voltage; gyrotrons; MIG; particle simulation; space charge effects

---

1. Introduction

Gyrotron is a kind of vacuum electronic device, of which the operation is based on the stimulated cyclotron radiation of electrons oscillating in a strong magnetic field. In a gyrotron, electrons that are emitted by the cathode, are accelerated in a strong electric field. While the electron beam travels through the intense magnetic field, the electrons start to gyrate at a specific frequency given by the strength of the magnetic field. In the beam–wave interaction circuit, located at the position with the highest magnetic field strength, the electromagnetic radiation is strongly excited. The output radiation leaves the gyrotron through a window and the spent electron beam is then dissipated in the collector. In general, the cyclotron interaction condition is $\omega - k_z v_z \cong s\Omega_e/\gamma$, where ω is the frequency of the electromagnetic wave, k_z is the axial wave number of the operating mode, v_z is the axial velocity of the electron passing through the cavity, s is the harmonic number, $\Omega_e(= eB_0/m_e c)$ is the rest-mass (m_e) electron cyclotron frequency, and γ is the relativistic factor of the electron. As the above principle of stimulated cyclotron radiation, the electrons energy can be transferred to the fast-wave in the interaction circuit. Thus, it can be operated at high-order mode in the cavity, the dimensions of the interaction structure can be much larger compared to the wavelength of the radiation, which provides capability to generate extremely high-power radiation.

Due to its excellent power output in the millimeter-wave and sub-millimeter wavebands, gyrotron has caused extensive and in-depth research by experts and scholars all over the world [1]. Lots of researches are focused on high-power, high-voltage gyrotrons used in ITER (International Thermonuclear Experimental Reactor), EAST(Experimental Advanced Superconducting Tokamak), etc. [2]. Currently, low-voltage gyrotron with hundreds of watts to several kilowatts output power has been arousing the interest of many scientists, because they are preferable from the engineering and reliability point of view [3,4]. For low-voltage gyrotrons, the problem of efficiency must be solved due to the weak relativistic factor. A prodigious amount of work has been done to improve the efficiency of gyrotron, like installing depressed collectors [5], changing the interaction structure [6], and using double electron-beam [7,8]. In this paper, we proposed a way to improve the beam–wave interaction by increasing the pitch factor. The interaction process of the gyrotron is mainly between the transverse electron cyclotron velocity and the perpendicular electric field, so increasing the velocity ratio can directly improve the interaction efficiency and thus becomes a key method to improve the efficiency of the low-voltage gyrotron. The efficiency of interaction is limited by the $\eta_e = E_\perp / E_{Total} \leq \alpha^2 / \left(1 + \alpha^2\right)$, where $E_\perp$ is the transverse energy, E_{Total} is the total beam energy, $\alpha = V_\perp / V_z$. Electron efficiency η_e was shown in Figure 1, when the velocity ratio reaches 3 the electron efficiency comes to 90% which is remarkable during the beam–wave interaction [9]. This scheme adopts a triode type magnetron injection gun with thermionic cathodes within the velocity spread under 10% [8]. The final ratio of transverse to longitudinal velocity in the interaction region is typically between $\alpha = 1$ and $\alpha = 2$ for gyrotrons [10]. The most frequent velocity ratio used in the MIG is 1.5 and it rarely exceeds 2, so the feasibility of ultra-high velocity ratio (MIG) is particularly valuable in compact gyrotron. In 2007, MIT has achieved a 3.5 kV gyrotron with a MIG, velocity ratio ranges from 2 and 5 with an operating current 10 mA [11]. The present development status of low-voltage (less than 10 kV) gyrotron improves the efficiency by increasing the pitch factor within low velocity spread, which are shown in Table 1. However, the operating current is often extremely low around 100 mA, with an output power of 10 W.

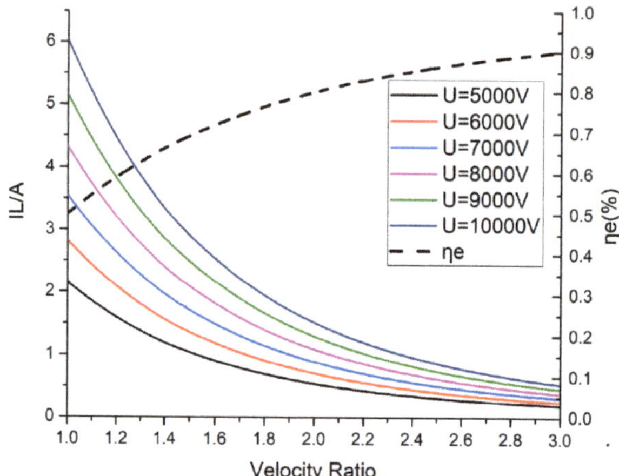

Figure 1. Effect of the velocity ratio variation on limiting current with different beam voltage.

Table 1. Performance parameters of latest low-voltage gyrotron.

Institution	Voltage	Current	Pitch Factor(α)	Velocity Spread	Power	Efficiency
MIT [11–15]	3.5 kV	50 mA	2	8%	12 W	6.8%
	12.3 kV	25 mA	No Mention		14 W	4%
	13 kV	100 mA	2	4%	16 W	1.2%
	10.1 kV	190 mA	2	4%	18 W	0.9%
	16.65 kV	110 mA	1.8	3%	9.3 W	0.5%
FIR FU [16]	10 kV	50 mA	No Mention		10 W	2%

MIT (Massachusetts Institute of Technology)
FIR FU (Research Center for Development of Far-Infrared Region, University of Fukui)

The required properties of the electron beam are basically determined by the chosen operating mode, frequency, and output power [3,17]. Parameters of this specific low-voltage gyrotron are shown in Table 2. According to the theory of space charge effects, the operating current is restricted. At the same time, considering the power requirements, beam voltage and current should reach a certain level. Restrictions of velocity ratio are investigated including the electrostatic field, limiting current, and velocity spread. Both theoretical analysis and particle-in-cell (PIC) simulation are adopted to investigate the feasibility of ultra-high velocity ratio MIG and give a specific design of MIG. The purpose of the present paper is to analyze the possibilities of ultra-high pitch-factor (α) for MIG operating at low-voltage with operating current about 500 mA. This paper is organized as follows. In Section 2, fundamental effects electrostatic and magnetostatic field on the α are analyzed theoretically, space charge effects in low-voltage operation are demonstrated, and velocity spread relevant to pitch factor is studied. In Section 3, a numerical simulation is implemented to obtain a specific MIG with ultra-high α and appropriate velocity spread. Variation of α and perpendicular velocity spread ratio with operating current and modulating voltage are studied. Conclusions and plans are discussed in Section 4.

Table 2. Basic specifications of low-voltage gyrotron.

Parameters	Value
Operating Mode	TE01
Beam Voltage (Ub)	10 kV
Beam Current (Ib)	0.5 A
Output Power (P)	0.5 kW
Output Frequency (f)	75 GHz
Main Magnet (B0)	2.7 T

2. Theoretical Calculation

2.1. Sensitivity Analysis of α for Axisymmetric Electric and Magnetic Field

The MIG is immersed in the crossed uniform electrostatic field and magnetic field. The electric field distribution is mainly determined by the geometry, voltage of the cathode, and modulating anode. The magnetic field is a uniform axial magnetic field that can ignore the radial component. Electrons moving in the increasing magnetic field leads to the momentum conversion to the transverse direction which is called adiabatic compression. According to adiabatic approximation theory, the transverse velocity at the ending position of the MIG area $\beta_{\perp 0}$ is defined by [18]:

$$\beta_{\perp 0} \approx \frac{1}{\gamma_0 c} b^{1/2} \frac{E_c \cos \theta_c}{B_c} = \frac{1}{\gamma_0 c} \frac{B_0^{1/2}}{B_c^{3/2}} E_c \cos \theta_c \tag{1}$$

b is the magnetic compression ratio $b = B_0/B_c$; B_0 and B_c are main magnetic field and cathode magnetic field, respectively. Magnetic compression could not be too high in case electrons reverse. γ_0 is the relativistic factor within the beam voltage in the absence of voltage depression, E_c is defined by

$$E_c \approx U_{\text{mod}} \frac{\cos\theta_c}{R_c \ln[1 + (d\cos\theta_c)R_c]} \tag{2}$$

d, R_c, θ_c are geometric factors which are shown in Figure 2, the MIG's Schematic. U_{mod} is the modulating anode voltage. So the velocity ratio could be defined by:

$$\alpha = \frac{\beta_{\perp 0}}{(\beta_0^2 - \beta_{\perp 0}^2)^{1/2}} = \frac{\beta_{\perp 0}}{(1 - \gamma_0^{-2} - \beta_{\perp 0}^2)} \tag{3}$$

where β_0 and γ_0 are the normal velocity and constant relativistic factors that are determined by the accelerating voltage Ub. Actually, the beam voltage has little influence on the α, and α can increase dramatically with the modulating anode voltage. Based on these equations, the pitch factor could be higher enough; however, Equations (1)–(3) have not taken the space charge effect into account. So the trade-off about these parameters is significant to investigate.

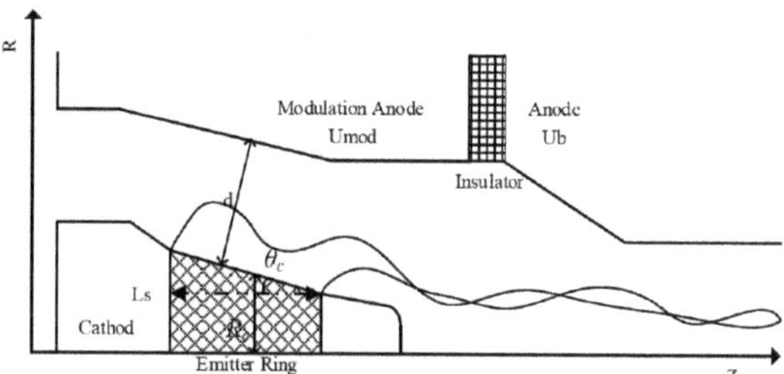

Figure 2. Schematic outline of the MIG.

2.2. Space Charge Limits

The space charge effect of the gyrating beam is defined in terms of the voltage depression V_{dep} and the limiting current IL [19,20]. The potential within the electron beam is reduced with respect to the wall potential due to the space charge in the electron beam. In a cylindrical waveguide, the depression voltage is the potential between the waveguide wall and the electron beam axis.

$$V_{\text{dep}} = \frac{1}{4\pi\varepsilon_0 c} \frac{I_b}{\beta_{\parallel}} G(R_i, R_o, R_a) \tag{4}$$

where the $G(R_i, R_o, R_a)$ is called the geometrical factor defined by:

$$G(R_i, R_o, R_a) = 2\ln\left(\frac{R_a}{R_o}\right) + \left[1 - \frac{2R_i^2}{(R_o - R_i)(R_o + R_i)} \ln\left(\frac{R_o}{R_i}\right)\right] \tag{5}$$

where R_a, R_o, and R_i are the radius of the interaction waveguide, the out-radius of the electron beam, and the inner-radius of the electron beam which are determined by the specific requirements of the interaction, $\beta_{\parallel}$ is the longitudinal velocity normalized to light velocity. The cross-section of the orbital cyclotron electron beam is shown in Figure 3. The average radius of guiding electron beam is defined by:

$$R_g = \frac{R_c}{b^{1/2}} \tag{6}$$

The R_c is the radius of the cathode. When a thin annular beam with the $R_o - R_i \ll R_i$, the geometrical factor can reduce to $G(R_i, R_o, R_a) \approx 2\, \ln(R_a / R_g)$, which results in:

$$V_{\text{dep}} \approx 60 \frac{I_b}{\beta_{\parallel}} \ln(\frac{R_a}{R_g}) \tag{7}$$

Based on the designed parameters, the geometrical factors are obtained. Coupling efficiency between the electron beam radius R_g and desired modes, which is satisfied as followed:

$$C_{mn} = (\frac{s^s}{2^s s!})^2 \frac{J_{m \pm s}^2(k_{mn} R_g)}{(x_{mn}^2 - m^2) J_m^2(x_{mn})} \tag{8}$$

where s is the harmonic order, k_{mn} is the eigenvalue of TE_{mn} mode, $x_{mn} = k_{mn}/R_a$.

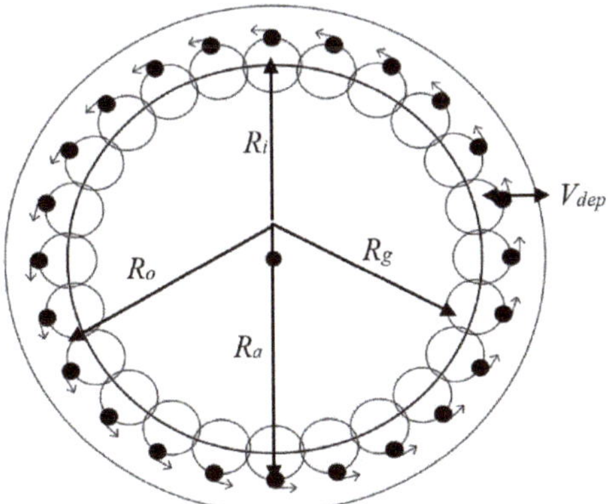

Figure 3. Schematic view of the electron beam cross-section at an arbitrary axial position.

The coupling efficiency of adjacent modes were shown in Figure 4, the guiding radius of 1.2 mm was chosen as the operating electron beam radius. In addition, the main electron beam factors are exhibited in Table 3 [19].

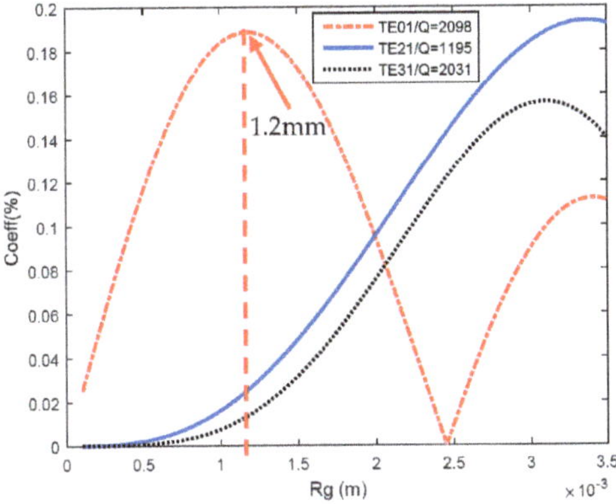

Figure 4. The coupling factor of different modes variation on beam radius.

Table 3. Cross-section geometrical parameters of electron beams.

Parameters	Value
The guiding radius (R_g)	1.2 mm
The inner radius (R_i)	1.1 mm
The outer radius (R_o)	1.3 mm
The interaction radius (R_a)	2.45 mm

The limiting current is defined by [21]:

$$IL = 1.707e4\gamma_0[1 - (1 - \beta_{\|0}^2)^{1/3}]^{1/2} \frac{1}{G(R_i, R_o, R_a)} \tag{9}$$

where $\beta_{\|0}$ is the axial propagation constant defined by $\beta_{\|0} = V_{z0}/c$, and G is the geometric factor defined previously. Results obtained from the equation IL are shown in Figure 1, there is a clear trend of decreasing limiting current when increasing α and dropping beam voltage. The electron efficiency is proportional to the velocity ratio. The potential dropping caused by the space charge effect limits the current flow in space.

However, a higher α will cause the limit current to decrease, and when the operating current is close to the limit current, it will cause the reversal of the electron beam, which drastically affects the emission and interaction efficiency. It seems very critical to obtain an optimal α value through theoretical analysis. To obtain a satisfactory electron beam quality, the operating current cannot exceed 0.5 of the limit current [9,17,21].

2.3. Velocity Spread

The total energy of the electron beam is always the same as a consequence of energy conservation [22]. However, during the adiabatic process, the perpendicular and parallel energy are distributed in a different way, which results in the velocity spread. The velocity spread is defined by

$$\frac{\Delta\beta_\|}{\beta_\|} = \alpha^2 \frac{\Delta\beta_\perp}{\beta_\perp} \tag{10}$$

where $\Delta\beta$ means the difference of maximum and minimum velocity. The velocity spread consists of axial and perpendicular velocity spread, which could reduce the efficiency of the beam–wave interaction and limit the achievable velocity ratio [23]. The velocity spread is implied as $\delta_{\beta\perp} = \Delta\beta_\perp/\beta_\perp$. Generally speaking, there are a lot of effects causing velocity spread, including initial thermal velocity spread, the roughness of the electron-emitting surface, the non-uniform distribution of the electric and magnetic field, and the effect of space charge [20]. This work only focused on the last two items in principle. Due to the non-uniform electric and magnetic field, the velocity spread is given below [24]:

$$\delta_{\beta\perp} \approx \frac{2R_c \cos\theta_c}{d} \tag{11}$$

where d is the distance between the cathode bottom and modulating anode, and it is often a large parameter. So, $\delta_{\beta\perp}$ from (11) is usually small. Space charge effects have a great influence on the velocity spread, especially for large currents. However, the effect of space charge on the electron beam depends on the beam trajectory. According to research in [24,25], an electron-gun with quasi-laminar beams could diminish $\delta_{\beta\perp}$. Thus, this design would use a MIG with a large inclination of emission to obtain quasi-laminar beams. Furthermore, the max velocity ratio is constrained by velocity spread, which is defined by:

$$\alpha_{max} = \frac{\overline{\beta_\perp}}{\beta_\|} = \frac{1}{(2\Delta\beta_{\perp max}/\overline{\beta_\perp})^{1/2}} \tag{12}$$

where $\bar{\beta}$ is the average velocity of electron beam. According to this equation, when the velocity spread is under 10%, the velocity ratio can reach 2.5. When the operating current is approaching the current limits electron beam begins to reverse, which causes the velocity spread sharply increased [26].

2.4. Summary

Taken together, these results suggest there is an indication that the high velocity ratio $\alpha \geq 2$ is fully achievable when the operating voltage and current are 10 kV and 0.5 A, respectively. Optimal electron beam parameters by calculation are shown in Table 4. But there are still a lot of conflicting requirements, so numerical simulation is implemented to find the optimal trade-offs.

Table 4. Optimal electron beam parameters by theoretical calculation.

Parameters	Value
Max Velocity (αmax)	2.5
Velocity Spread ($\delta_{\beta\perp}$rms)	10%
Voltage Depression (Vdep)	286 V

3. Numerical Simulation by Magic 2D PIC Code

Based on the theory of the MIG [18,27], fundamental design factors were calculated and given in Table 5. Particle simulations are important to verify the beam performance of any specific gun design. 2D Magic code was adopted to optimize the parameters of the MIG and Magic was considered as a convinced software to calculate the particle motion and beam–wave interaction, which includes the space charge effect. The calculation in the Magic 2D code was performed with a mesh size of 0.3 and 0.1 mm in the z and r direction. The geometry model and electron trajectory cooperated with the axial magnetic field and were shown in Figure 5. Beam voltage and current were 10 kV and 460 mA, respectively. Electron momentum is defined by $P = m \times v$, perpendicular and longitudinal momentum were shown in Figure 6a,b. Energy transferred to the transverse direction from the axial direction due to the adiabatic compression. When the transverse-to-axial velocity ratio reached the maximum, as shown in Figure 6c, electrons leave the electron area and then turn into the interaction area. Specific parameters of the electron beam are shown in Table 6, with 2.55 velocity ratio and 9.33% perpendicular spread ratio.

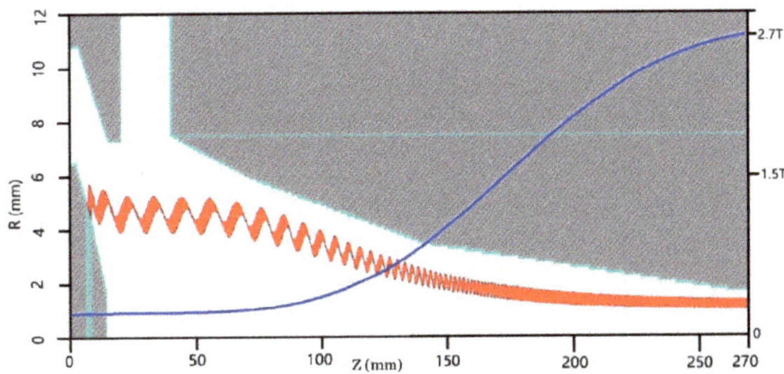

Figure 5. Optimized geometry, magnetic field profile and beam trajectory of MIG by Magic code.

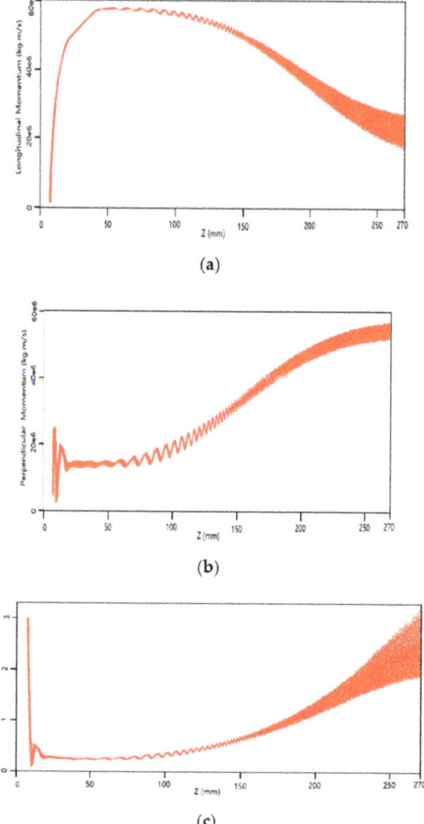

(a)

(b)

(c)

Figure 6. The value for (**a**) longitudinal momentum, (**b**) perpendicular momentum, and (**c**) transverse-to-axial velocity ratio as a function of axial distance.

Table 5. Optimal geometric factors of MIG.

Parameters	Value
Modulating Voltage (Umod)	7.5 kV
Average Radius of Cathode (Rc)	4.65 mm
Thickness of Emitter Ring (Ls)	1.5 mm
Tilt Angle of Cathode (θc)	18.4°
Distance between Cathode and Modulating Anode (d)	3.91 mm

Table 6. Optimized beam parameters obtained by Magic 2D simulation.

Parameters	Value
Average Beam Radius (Rg)	1.2 mm
Velocity Ratio (α)	2.55
Perpendicular Velocity Spread Ratio ($\delta_{\beta\perp rms}$)	9.33%
Full Radial Width	0.22 mm

Based on the theoretic calculation in Figure 1, when the limiting current is 0.824 A, the operating voltage is 10 kV and the velocity ratio is 2.55. Due to the working current being generally selected as half of the limit current, numerical simulation is adopted to investigate the sensitivity of the beam current around the 0.416 A for α and spread. From Figure 7 it can be seen that when the current is 0.416 A, the velocity ratio is the lowest, with velocity spread of 9.4% and α 2.55. This is also self-consistent with the previous maximum velocity ratio and velocity spread relationship in Section 2.3 Equation (12). As shown in Figure 7, with the increasing of the beam current, the velocity ratio would decrease, but when it exceeds a value, the velocity spread would boost rapidly.

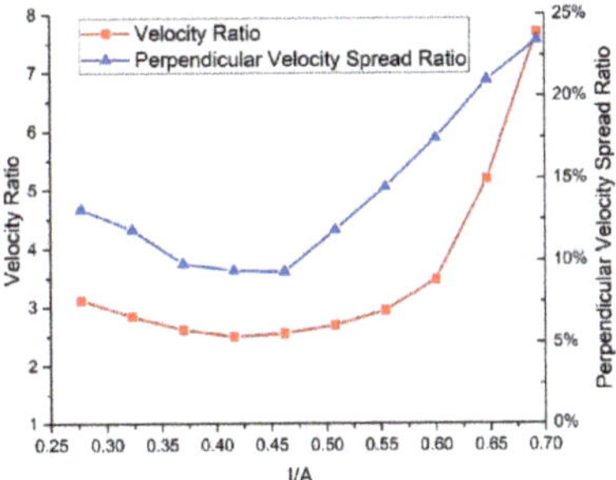

Figure 7. Sensitivity analysis of beam velocity ratio and perpendicular velocity spread ratio for modulating anode voltage.

Taking into consideration both the experiment and the theory, the electrical field of the cathode can be convenient to modulate to obtain a satisfactory electron beam. So, particle simulation is adopted to analyze the sensitivity of the modulating voltage for the velocity ratio and velocity spread when beam voltage and current are 10 kV and 0.46 A, respectively. What can be clearly seen in Figure 8 is the continual growth of the velocity ratio and slight fall of the perpendicular velocity spread with the increasing modulating voltage. Therefore, the simulation is consistent with the theory study.

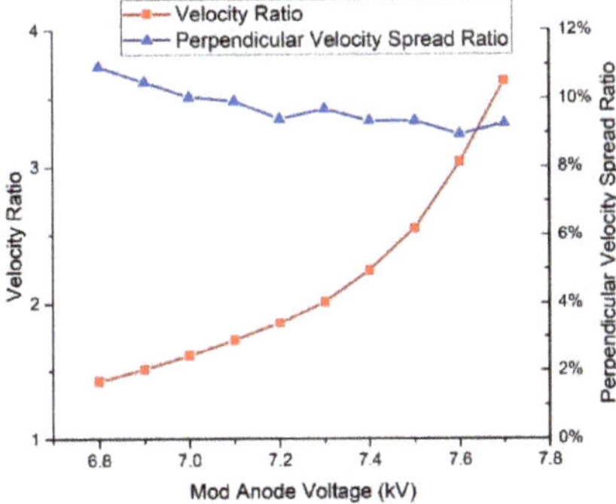

Figure 8. Variation of velocity ratio (α) and perpendicular velocity spread ratio with operating current.

4. Conclusions

The purpose of the current study was to determine the feasibility of the ultra-high velocity ratio of MIG for low-voltage gyrotron. The theory of space charge effects and particle simulation were adopted to investigate the velocity ratio and velocity spread of MIG.

This work proposes a direct and convenient way (ultra-high velocity ratio) to enhance the efficiency of low-voltage gyrotron, which operates at 10 kV or even low-voltage, while maintaining a large operating current (about 0.5 A), improving the velocity ratio, and restricting the velocity spread. Furthermore, the theoretical analysis and numerical simulation were adopted to verify the feasibility of the ultra-high velocity ratio. The results show that the velocity ratio of 2.55 and the velocity spread of 9.33% are achievable by optimizing the electron gun geometric factors, magnetic field, operating current, and modulating anode voltage.

Author Contributions: D.L. and W.F. contributed to the overall study design, analysis, computer simulation, and writing of the manuscript. X.G., T.Y., C.Z., C.C., M.H., and Y.Y. provided technical support and revised the manuscript. All authors have read and agreed to the published version of the manuscript.

Funding: This work was supported by the National Key Research and Development Program of China under 2019YFA0210202, the National Natural Science Foundation of China under Grant 61971097, the Sichuan Science and Technology Program under Grant 2018HH0136, and the Terahertz Science and Technology Key Laboratory of Sichuan Province Foundation under Grant THZSC201801.

Conflicts of Interest: The authors declare no conflict of interest.

References

1. Thumm, M. State-of-the-art of high-power gyro-devices and free electron masers. *J. Infrared Millim. Terahertz Waves* **2020**, *41*, 1–140.

2. Kartikeyan, M.V. Gyrotrons-high-power microwave and millimeter wave technology. In *Advanced Texts in Physics*; Springer Scientific: Berlin, Germany, 2004.

3. Bratman, V.L.; Fedotov, A.E.; Kalynov, Y.K.; Makhalov, P.B.; Osharin, I.V. Numerical Study of a Low-Voltage Gyrotron ("Gyrotrino") for DNP/NMR Spectroscopy. *IEEE Trans. Plasma Sci.* **2017**, *45*, 644–648. [CrossRef]

4. Glyavin, M.Y.; Zavolskiy, N.A.; Sedov, A.S.; Nusinovich, G.S. Low-voltage gyrotrons. *Phys. Plasm.* **2013**, *20*, 033103. [CrossRef]

5. Manuilov, V.N.; Morozkin, M.V.; Luksha, O.I.; Glyavin, M.Y. Gyrotron collector systems: Types and capabilities. *Infrared Phys. Technol.* **2018**, *91*, 46–54. [CrossRef]

6. Kishko, S.A.; Ponomarenko, S.S.; Kuleshov, A.N.; Zavertanniy, V.V.; Yefimov, B.P.; Alexeff, I. Low-voltage cyclotron resonance maser. *IEEE Trans. Plasma Sci.* **2013**, *41*, 2475–2479. [CrossRef]

7. Kuleshov, A.; Ishikawa, Y.; Tatematsu, Y.; Mitsudo, S.; Idehara, T.; Khutoryan, E.; Kishko, S.; Ponomarenko, S.; Glyavin, M.; Bandurkin, I.; et al. Low-voltage operation of the double-beam gyrotron at 400 GHz. *IEEE Trans. Electron Devices* **2020**, *67*, 673–676. [CrossRef]

8. Manuilov, V.N.; Tsvetkov, A.I.; Glyavin, M.Y.; Mitsudo, S.; Idehara, T.; Zotova, I.V. Universal electron gun design for a CW third harmonic gyrotron with an operating frequency over 1 THz. *J. Infrared Millim. Terahertz Waves* **2020**, *41*, 8–9. [CrossRef]

9. Nusinnovich, G.S. *Introduction to the Physics of Gyrotrons*; Johns Hopkins Univercity Press: Barltimore, MD, USA, 2012.

10. Kreischer, K.E.; Kimura, T.; Danly, B.G.; Temkin, R.J. High-power operation of a 170 GHz megawatt gyrotron. *Phys. Plasmas* **1997**, *4*, 1907–1914. [CrossRef]

11. Hornstein, M.K.; Bajaj, V.S.; Griffin, R.G.; Temkin, R.J. Efficient low-voltage operation of a CW gyrotron oscillator at 233 GHz. *IEEE Trans. Plasma Sci.* **2007**, *35*, 27–30. [CrossRef]

12. Joye, C.D.; Griffin, R.G.; Hornstein, M.K.; Hu, K.N.; Kreischer, K.E.; Rosay, M.; Shapiro, M.A.; Sirigiri, J.R.; Temkin, R.J.; Woskov, P.P. Operational characteristics of a 14-W 140-GHz gyrotron for dynamic nuclear polarization. *IEEE Trans. Plasma Sci.* **2006**, *34*, 518. [CrossRef]

13. Torrezan, A.C.; Han, S.; Mastovsky, I.; Shapiro, M.A.; Sirigiri, J.R.; Temkin, R.J.; Barnes, A.B.; Griffin, R.G. Continuous-Wave Operation of a Frequency-Tunable 460-GHz Second-Harmonic Gyrotron for Enhanced Nuclear Magnetic Resonance. *IEEE Trans. Plasma Sci.* **2010**, *38*, 1150–1159. [CrossRef] [PubMed]

14. Torrezan, A.C.; Shapiro, M.A.; Sirigiri, J.R.; Temkin, R.J.; Griffin, R.G. Operation of a continuously frequency-tunable second-harmonic CW 330-GHz gyrotron for dynamic nuclear polarization. *IEEE Trans. Electron Devices* **2011**, *58*, 2777–2783. [CrossRef]

15. Jawla, S.K.; Griffin, R.G.; Mastovsky, I.A.; Shapiro, M.A.; Temkin, R.J. Second harmonic 527-GHz gyrotron for DNP-NMR: Design and experimental results. *IEEE Trans. Electron Devices* **2019**, *99*, 1–7. [CrossRef] [PubMed]

16. Ausu, L.; Idehara, T.; Ogawa, I. Design of a compact CW THz gyrotron. In Proceedings of the 34th International Conference Infrared, Millimeter, Terahertz Waves, Busan, Korea, 21–25 September 2009; pp. 1–2.

17. Kumar, N.; Singh, U.; Khatun, H.; Kumar, A.; Singh, T.P.; Sinha, A.K. *Sensitivity Analysis of Electron Beam Velocity Ratio for 42 GHz, 200 kW Gyrotron*; Frequenz: Berlin, Germany, 2010; Volume 64, pp. 93–96.

18. Edgcombe. *Gyrotron Oscillators: Their Principles and Practice*; Taylor and Francis: London, UK, 1993.

19. Drobot, A.T.; Kim, K. Space charge effects on the equilibrium of guided electron flow with gyromotion. *Int. J. Electron.* **1981**, *51*, 351–367. [CrossRef]

20. Gilmour, A.S.; Ebrary, I. *Klystrons, Traveling Wave Tubes, Magnetrons, Cross-Field Amplifiers, and Gyrotrons*; Artech House: Norwood, MA, USA, 2011.

21. Ganguly, A.K.; Chu, K.R. Limiting current in gyrotrons. *Int. J. Infrared Millim. Waves* **1984**, *5*, 103–121. [CrossRef]

22. Singh, A.; Jain, P.K. Effects of electron beam parameters and velocity spread on radio frequency output of a photonic band gap cavity gyrotron oscillator. *Phys. Plasmas* **2015**, *22*, 3945. [CrossRef]

23. Fokin, A.P.; Glyavin, M.Y.; Nusinovich, G.S. Effect of ion compensation of the beam space charge on gyrotron operation. *Phys. Plasmas* **2015**, *22*, 351. [CrossRef]

24. Tsimring, S.E. Gyrotron electron beams: Velocity and energy spread and beam instabilities. *Int. J. Infrared Millim. Waves* **2001**, *22*, 1433–1468. [CrossRef]

25. Kuftin, A.N.; Lygin, V.K.; Tsimring, S.E.; Zapevalov, V.E. Numerical simulation and experimental study of magnetron-injection guns for powerful short-wave gyrotrons. *Int. J. Electron.* **1992**, *5*, 1145–1151. [CrossRef]

26. Nguyen, K.T.; Danly, B.G.; Levush, B.; Blank, M.; True, R.; Good, G.R.; Hargreaves, T.A.; Felch, K.; Borchard, P. Electron gun and collector design for 94-GHz gyro-amplifiers. *IEEE Trans. Plasma Sci.* **1998**, *26*, 799–813. [CrossRef]

27. Fu, W.; Yan, Y.; Liu, S. Study on a 60 kV/5 A magnetron injection gun for 200 GHz electron cyclotron master. *Front. Electr. Electron. Eng. China* **2009**, *4*, 440. [CrossRef]

electronics

MDPI

Article

A Compact Modular 5 GW Pulse PFN-Marx Generator for Driving HPM Source

Haoran Zhang [1,2], Ting Shu [1,2], Shifei Liu [1,2], Zicheng Zhang [1,2,*], Lili Song [1,2,*] and Heng Zhang [1,2]

[1] College of Advanced Interdisciplinary Studies, National University of Defense Technology, Changsha 410073, China; hrzhang@nudt.edu.cn (H.Z.); ting.shu@nudt.edu.cn (T.S.); liushifei16@nudt.edu.cn (S.L.); zhnudt@126.com (H.Z.)

[2] State key Laboratory of Pulsed Power Laser Technology, Changsha 410073, China

[*] Correspondence: zczhang@nudt.edu.cn (Z.Z.); phoezy@126.com (L.S.)

Abstract: A compact and modular pulse forming network (PFN)-Marx generator with output parameters of 5 GW, 500 kV, and 30 Hz repetition is designed and constructed to produce intense electron beams for the purpose of high-power microwave (HPM) generation in the paper. The PFN-Marx is composed by 22 stages of PFN modules, and each module is formed by three mica capacitors (6 nF/50 kV) connected in parallel. Benefiting from the utilization of mica capacitors with high energy density and a mini-trigger source integrated into the magnetic transformer and the magnetic switch, the compactness of the PFN-Marx system is improved significantly. The structure of the PFN module, the gas switch unit, and the connection between PFN modules and switches are well designed for modular realization. Experimental results show that this generator can deliver electrical pulses with the pulse width of 100 ns and amplitude of 500 kV on a 59-ohm water load at a repetition rate of 30 Hz in burst mode. The PFN-Marx generator is fitted into a cuboid stainless steel case with the length of 80 cm. The ratio of storage energy to volume and the ratio of power to weight of the PFN-Marx generator are calculated to be 6.5 J/L and 90 MW/kg, respectively. Furthermore, utilizing the generator to drive the transit time oscillator (TTO) at a voltage level of 450 kV, a 100 MW microwave pulse with the pulse width of 20 ns is generated.

Keywords: PFN-Marx; compact; modular; trigger source; gas switch; mica capacitor

Citation: Zhang, H.; Shu, T.; Liu, S.; Zhang, Z.; Song, L.; Zhang, H. A Compact Modular 5 GW Pulse PFN-Marx Generator for Driving HPM Source. *Electronics* **2021**, *10*, 545. https://doi.org/10.3390/electronics10050545

Academic Editor: Mikhail Glyavin

Received: 31 January 2021
Accepted: 21 February 2021
Published: 26 February 2021

Publisher's Note: MDPI stays neutral with regard to jurisdictional claims in published maps and institutional affiliations.

1. Introduction

Intense electron beams have a wide range of applications in science research and industry fields, such as in the high-power microwave (HPM) and flash X-ray [1–3]. Producing high voltage (with several hundred kilovolts), rectangular pulses are one of the crucial parts. Recently, with the development of the energy storage technology, pulsed power technology has been developed in the direction of compactness, miniaturization, and light weight. Because the pulse forming network (PFN)-Marx generator has the natural advantage of the integration of pulse modulation and pulse voltage accumulation, it has been developed rapidly in recent years and is regarded as one of the most promising types for compactness, modularization, light weight, and miniaturization realization [4–9]. In this paper, two key points, including the compactness and modularization, of the PFN-Marx generator are discussed in detail.

Firstly, the literature on the research of the Marx generator over the past few decades has been reviewed. Institutes around the world, including the French-German Research Institute of Saint-Louis (ISL), Texas Technology University (TTU), Commissariat à l'Energie Atomique et aux Energies Alternatives (CEAE), Applied Physical Electronics L. C. (APELC), etc., have developed a series of compact Marx generators with different circuit topologies and the typical parameters of the generators are listed in Table 1 [10–15]. Generally, the ratio of storage energy to volume (E/V) and the ratio of power to weight (P/W) are calculated to describe the compactness of the generator. According to the limited parameters of the generator in the literature, the E/V and P/W are calculated to several J/L and dozens of MW/kg, respectively. Actually, a variety of methods, including structure optimization, novel topology circuits, large energy density capacitors, etc., have been studied by researchers to realize the compact design. However, the physical limits of insulation are roughly the same in spite of the circuit topologies and structures. The effective way to improve the compactness of the generator is by using the high energy density capacitors. Generally, the ceramic capacitors and the film capacitors are more popular in Marx generator applications. The ceramic capacitors have the advantages of small size and low self-inductance, and the film capacitors have the advantages of large capacitance and high withstand voltage. In terms of compact Marx generators, ceramic capacitors are used more often because of their small size. However, the energy density of ceramic capacitors is low because it is difficult to balance the large capacitance, high withstand voltage, and small volume. Recently, mica capacitors have gradually drawn researchers' attention because of their good electric characteristics [11]. Mica capacitors have the advantages of large capacitance, high withstand voltage, and small size, which make them a promising alternative to ceramic capacitors in the compact Marx generators.

Table 1. The parameters of the typical compact pulse forming network (PFN)-Marx generators.

Research Institute	Output Power	Output Voltage	Pulse Width	E/V	P/W
ISL	2.3 GW	342 kV	60 ns	2 J/L	70 MW/kg
TTU	2.2 GW	210 kV	200 ns	9 J/L	-
CEAE	1.6 GW	400 kV	85 ns	2 J/L	36 MW/kg
APELC	5 GW	350 kV	200 ns	2 J/L	-

Secondly, the modular design of the PFN-Marx generator generally contains two meanings. One is the modular implementation of the PFN-Marx body. Obviously, a variety of PFN-Marx generators with different structures have been designed. However, once the generator is assembled, the PFN module is difficult to replace. The modular design requires that the PFN module can be taken out and replaced from the system individually and therefore the PFN-Marx generator can be convenient for maintenance and parameter adjustment. In this case, the optimization of the PFN module design and assembling method becomes necessary. On the other hand, the PFN-Marx generator system includes many subsystem modules by their functions, which are the primary energy module, PFN-Marx body, trigger source, and vacuum diode. In order to realize a compact PFN-Marx generator system, the design of the subsystem modules should be optimized as well.

In this paper, mica capacitors with large energy density are assembled in the PFN-Marx generator to improve its compactness. The PFN module and the integrated gas gap switch are well designed for modular realization. In Section 2, the design and the assembly of the PFN module are elaborated in detail and the subsystem module of the PFN-Marx system is introduced in Section 3, followed by experimental results and a summary in Sections 4 and 5, respectively.

2. Modular PFN-Marx Design

2.1. Overall Description

The PFN-Marx generator is designed for driving a HPM source. In order to make the generator have a wider range of application, the peak power (P) and the load impedance (R) are designed as 5 GW and 50–60 Ω, respectively.

Figure 1 shows the schematic circuit of the PFN-Marx. The type-E is applied in this generator, because the equal L-C sections are easy for modular realization and fabrication in practice. The established parameters of the PFN-Marx are listed in Table 2. The number of the node capacitor n and the number of the PFN module m can be calculated to 3 and 22, respectively, by Equations (1)–(3).

$$\tau = 2n\sqrt{LC} \tag{1}$$

$$R = m\sqrt{\frac{L}{C}} \tag{2}$$

$$V_{output} = \frac{m V_{ch}}{2} \tag{3}$$

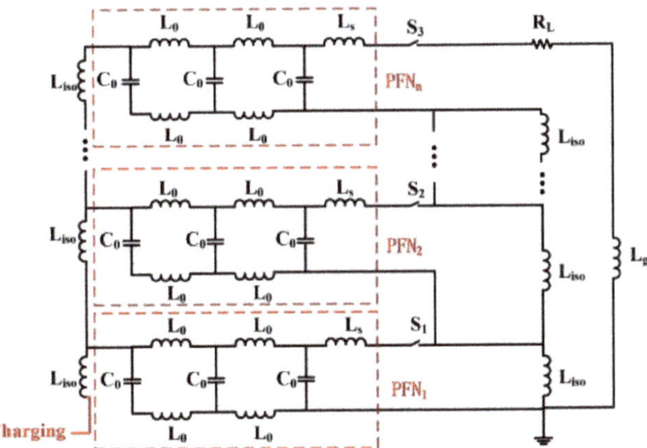

Figure 1. Schematic circuit of the PFN-Marx.

Table 2. PFN-Marx generator specifications.

Symbol	Description	Value
P	Peak power	5 GW
V_{load}	Load voltage	500 kV
R	Load impedance	50–60 Ω
T	Pulse width	90–100 ns
C_0	Node capacitor	6 nF
L_0	Node inductor	15 nH
V_{ch}	Charging voltage	50 kV
Rep	Repetition rate	30 Hz

In order to realize the repetition operation of the generator, the inductive isolation of the charging is used and the optimized isolation inductor L_{iso} is calculated to 20 μH. The inductance of the switch and the switch leading wire L_s is set as 45 nH. The simulation result is shown in Figure 2. The load voltage waveform characteristics include the rise-time of about 20 ns, pulse width of about 90 ns, flat-top of about 50 ns, and amplitude of about 530 kV with the charging voltage of 50 kV.

A series of methods are used for the realization of the modular PFN-Marx body, including the PFN module design, the connection method between the PFN module and the electrode of the switch, and the common switch housing. The key components of the modular PFN-Marx generator are illustrated as follows.

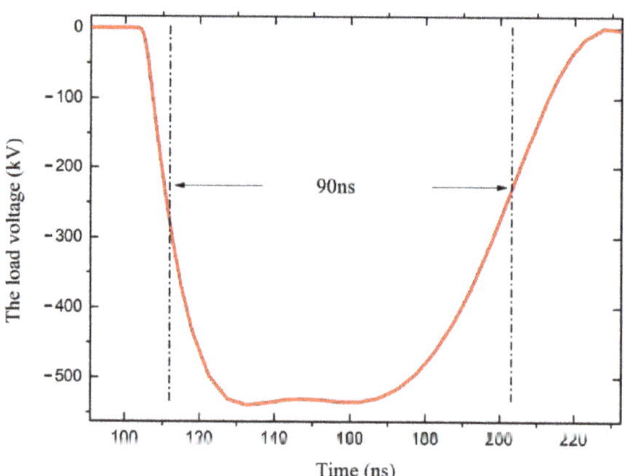

Figure 2. The load voltage waveform in simulation.

2.2. PFN Module

The assembly diagram of the PFN module is shown in Figure 3. Each PFN module consists of three 6000 pF ($22 \times 40 \times 95$ mm^3) mica capacitors connected in parallel by means of copper strips, and the copper strips also serve as the node inductors. The mica capacitors are inserted into a nylon case with external dimensions of 178 mm $\times$ 150 mm $\times$ 30 mm. The energy density of the mica capacitor is nearly 90 J/L, which is over three times that of the commercial ceramic capacitor with 2000 pF (φ60 mm $\times$ 32 mm). The selection of the mica capacitor with large energy density sets a good foundation for the compact design. Each mica capacitor assembly was tested under the DC voltage of 50 kV with a time duration of 1 min before usage. Then, the connected mica capacitors were inserted into a nylon box to ensure the insulation distance between the adjacent modules and to provide mechanical support. Finally, the key point of the modular realization is the connection between the PFN module and the switch. Here, the socket joint mode is adopted in which two plug electrodes are welded with the PFN via two high-voltage wires. In this way, the PFN module can be taken out individually from the PFN-Marx body and the structure of the PFN-Marx remains unchanged.

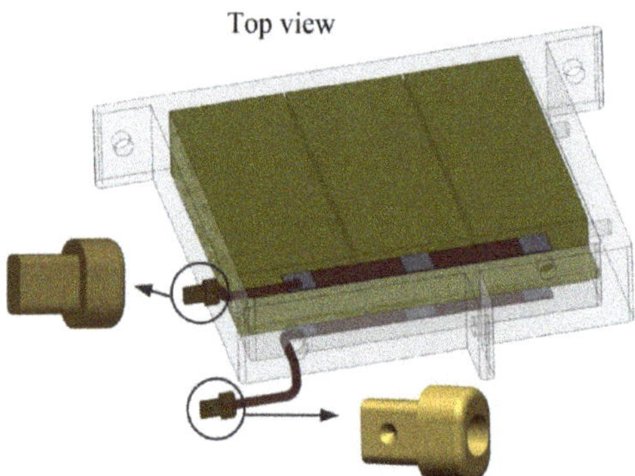

Figure 3. The assembly diagram of the PFN module.

2.3. Gas Gap Switches

Switches, as energy-shifting elements in the pulse power system, play an important role in the pulse compression and power enhancement. Generally, the generation of high voltage of the PFN-Marx generator is that the PFN modules are erected via the Marx-type method. Therefore, the number of switches of the PFN-Marx generator is the same as the number of the PFN modules. For a PFN-Marx generator, the operating characteristics of switches are related to the wave erection. So, the switches are designed specifically to ensure the performance and at the same time to realize the modular function of the PFN-Marx generator. In practice, all of the switch electrodes of the generator are assembled in a nylon housing with external dimensions of 695 mm × 85 mm × 60 mm, as shown in Figure 4. Actually, there are two main advantages of the integrated switch design: one is that the gas pressure of the switch can be adjusted independently, so that the operation voltage of the switch can be adjusted in a wide range; the other advantage is that the ultraviolet light, which is generated when the first few stages of the switches close, facilitates the switching of the later ones. In this way, the erected time of the PFN-Marx can be decreased.

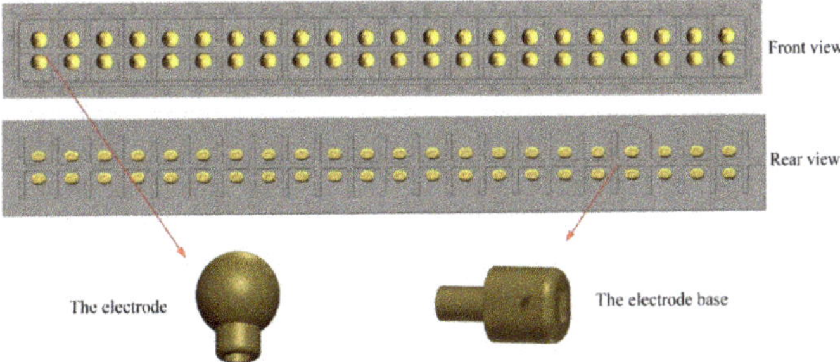

Figure 4. The assembly diagram of the gas gap switches.

The electrode of the switch has a spherical geometry with the material of copper. The diameter of the electrode is 15 mm and the gap distance is 10 mm. These spherical electrodes are assembled with the electrode bases, as shown in Figure 4. A rectangular countersink hole is milled at the end of the electrode base to assemble the plug electrode, which is shown in Figure 3. The plug electrode and the electrode base are secured by a metal screw. Therefore, the switch of the generator is an independent assembly and the PFN module can realize the quick dismantling function.

In fact, in order to improve the operation stability of the generator, the switch is normally triggered by a high-voltage electric pulse. The switches of the first two stages are a three-electrode configuration constructed by a 2 mm stainless steel needle into a 15 mm brass sphere electrode, and Figure 5 shows the schematic diagram of the three-electrode switch. The gap length between triggered electrode and main electrode and between two main electrodes is 3 mm and 10 mm, respectively. The dielectric material is assembled in the electrode to insulate the trigger electrode. The lifetime of the single triggered switch was tested in the experiment. The insulation gas is N_2 mixed with 15% SF_6 and the gas pressure in the switch house is 80 kPa. The switch was tested in 10 Hz repetition mode with an operation voltage of 50 kV over 10,000 shots. Figure 6 shows the photograph of the triggered switch after 100,000 shots. This work indicates that the gas gap switch in Figure 4 would have a long lifetime when it periodically changes the working gas.

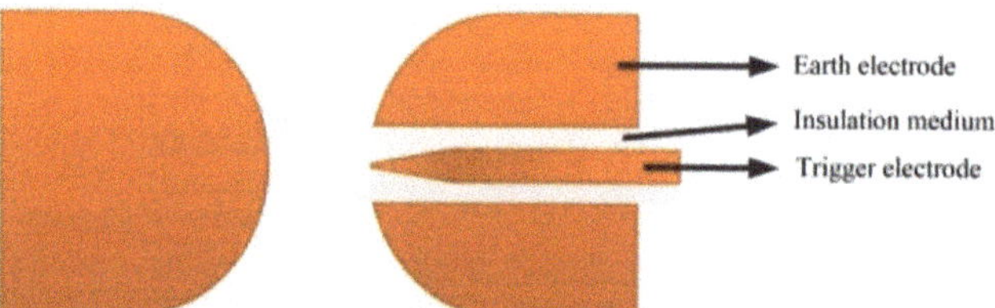

Figure 5. Schematic diagram of the trigger switch.

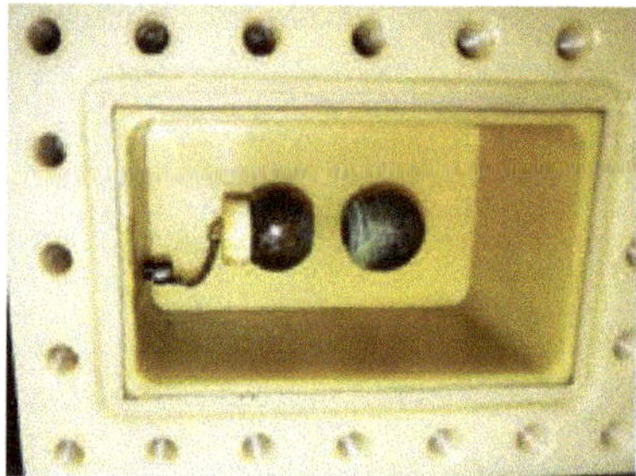

Figure 6. Photograph of the trigger switch.

2.4. The Modular PFN-Marx Generator Assembly

Twenty-two PFN modules are constructed in the form of a drawer-type geometry, which can minimize circuit inductance as well as realize the modular function, as shown in Figure 7. There are four components of the PFN-Marx body, including three support rods, PFN modules, and a switch assembly. Firstly, the PFN modules are assembled on nylon rods with nylon screws. Secondly, the switch assembly, the nylon rods, and the PFN modules are assembled on the nylon plate. Finally, the plug electrodes are inserted into the electrode bases of the switch and assembled with metal screws. In this way, the PFN module can be taken out individually and the PFN-Marx structure is kept unchanged. Therefore, the convenience of the PFN-Marx generator's maintenance can be improved significantly.

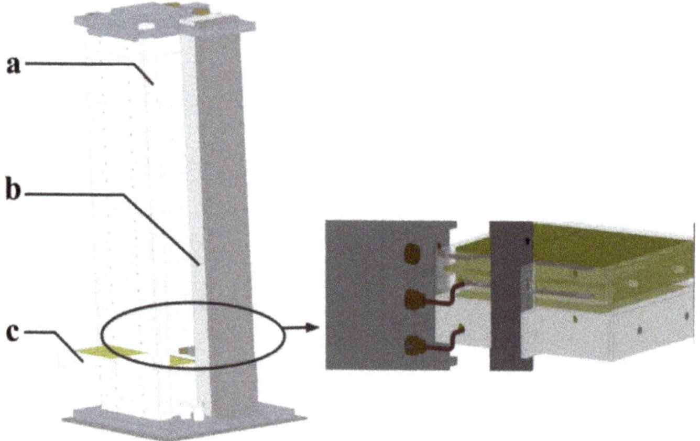

Figure 7. Assembly diagram of the PFN-Marx (a—support rod, b—gas switch, c—PFN module).

3. Experimental Setup

3.1. Primary Energy Subsystem

A primary energy subsystem was designed and established for repetition operation of the PFN-Marx generator, and the equivalent circuit is shown in Figure 8. At the beginning of the generator operation, the intermediate energy storage capacitor C_f and the primary capacitor of the pulse transformer C_p are charged by an AC transformer TF_1, and the maximum charging voltage is 2800 V. The mica capacitors in the PFN-Marx generator are charged by an air-core pulse transformer TF and the maximum charging voltage V_{ch} is up to 50 kV. The charging voltage waveform is shown in Figure 9.

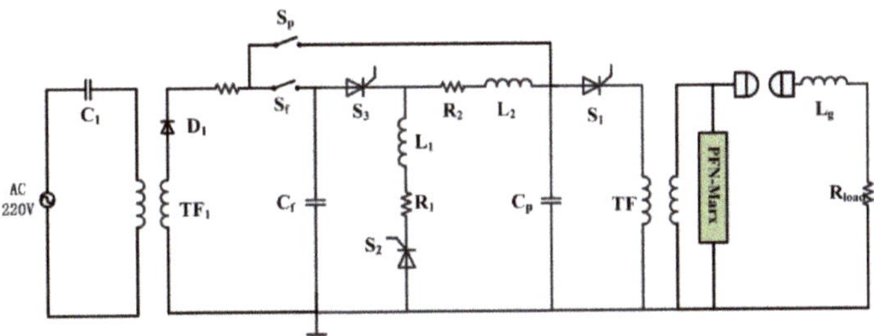

Figure 8. Schematic circuit of the primary energy subsystem.

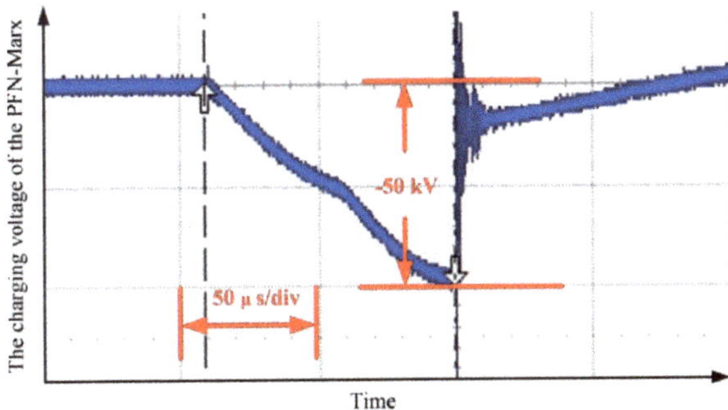

Figure 9. The charging voltage waveform of the PFN-Marx generator.

The working process of the system in the case of repetition mode is described as follows. The operation sequence of the thyristors is shown in Figure 10. Firstly, the C_f and C_p are charged to U_f and U_0, respectively. Next, when the thyristor S_1 is closed at time T_1, the C_p will discharge to the PFN-Marx through the TF. After that, the C_p has the negative residual voltage Urec because of the principle of the unidirectional current of the thyristor. Then, the thyristor S_2 is closed at time T_2, and the voltage polarity of the C_p is reversed via the loop of C_p-S_2-R_1-L_1-R_2-L_2-C_p. Finally, the thyristor S_3 is closed at time T_3, the C_p is charged to U_0 again, and the system waits for the next operation command. So, the key point of the repetition operation is that the control of the on time of the thyristors and the pulse number is related to the capacitance of the C_f.

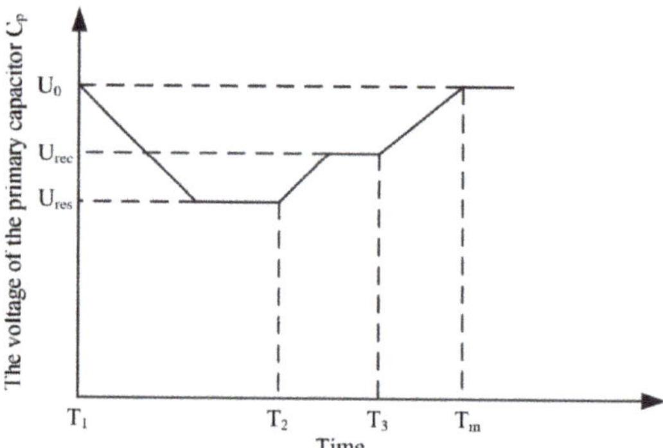

Figure 10. The sequence diagram of the thyristors.

3.2. Trigger Source

An important way of improving the synchronization performance of the switch is to reduce the rise-time of the trigger pulse and increase the trigger pulse amplitude. Here, a compact Marx generator, which is based on a magnetic switch, is designed for a trigger source [16]. The key point of this trigger source is that the integrated design of the transformer and the switch is realized, so that the volume of it can be reduced significantly. The basic circuit of the trigger source is shown in Figure 11. The magnetic switch is operated as a transformer during the charging process at first. When the magnetic core is saturated, the inductance of the secondary windings is decreased rapidly. At this time, the energy is sharply delivered to the load.

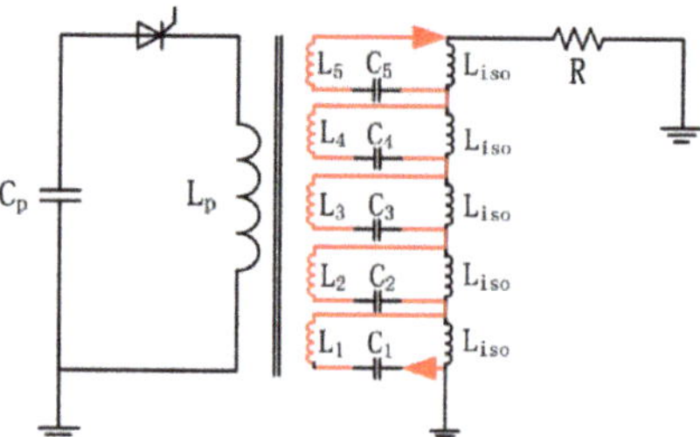

Figure 11. The circuit of the trigger source.

The trigger source assembly in the experiment is shown in Figure 12. It can be seen that the trigger source is very compact and its external dimensions are 150 mm × 200 mm × 50 mm. The Marx generator is assembled in a dielectric material house and pressurized SF_6 gas is used for electrical insulation. The essential characteristic of the magnetic switch is that the time of core saturation is consistent if the charging voltage of the primary capacitor is consistent. Therefore, the consistency of the trigger pulses is pretty good, as shown in Figure 13. The trigger source is operated in 30 Hz mode and the experimental results show that the amplitude of the trigger pulses is over 70 kV and the jitter is less than 10 ns. This trigger source provides a stable and reliable trigger pulse for the switches.

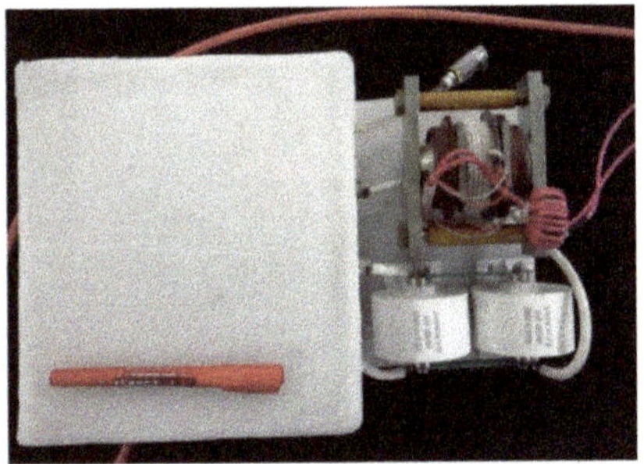

Figure 12. The photograph of the trigger source.

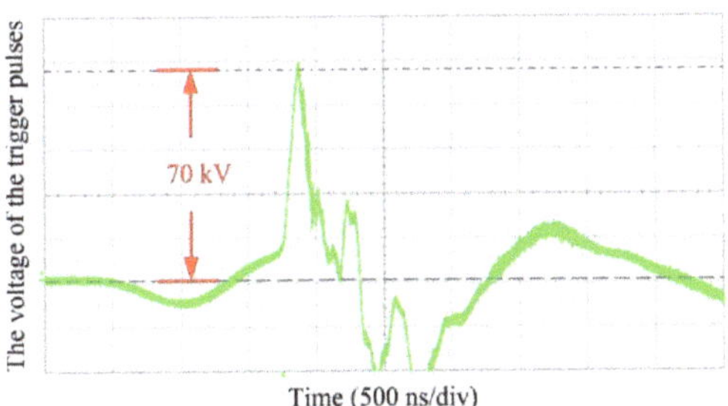

Figure 13. Screen shot of 20 pulses at 30 Hz operation of the trigger source.

3.3. Experimental Setup

In order to evaluate the performance of the compact modular PFN-Marx generator, the system of the generator was arranged as shown in Figure 14. The generator mainly consists of five parts, including a control subsystem, a trigger source, a primary energy subsystem, a PFN-Marx body, and a load. The charging voltage was measured with a commercial polestar probe with 2000:1. A water resister divider was inserted into the diode insulator to measure the load voltage. It can be seen in Figure 14 that the compactness assembly is accomplished. After the system is assembled in practice, the length of the PFN-Marx body is about 80 cm and its weight is about 56 kg.

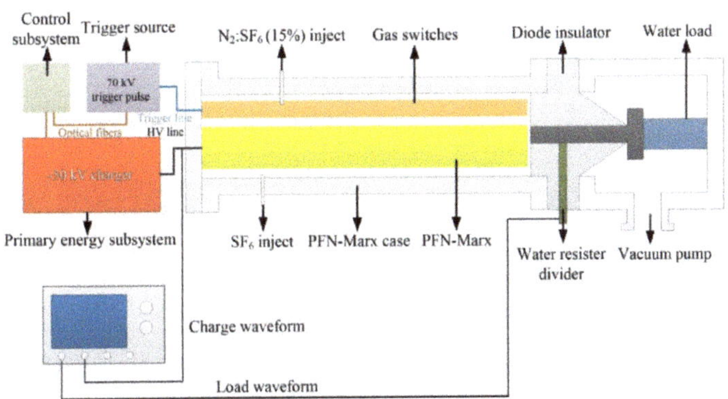

Figure 14. Schematic view of the PFN-Marx system.

4. Experimental Results

4.1. Experimental Results on a Water Load

Initial testing of the compact PFN-Marx generator involved firing the system at the charging voltage of 50 kV into a shot, and a 59 Ω coaxial $CuSO_4$ water load was connected to the generator to evaluate the operating characteristic of the generator. The water load was assembled in the vacuum diode as shown in Figure 14.

The generator was tested in single shot and burst mode, respectively. The pure SF_6 of pressure of 180 kPa was injected into the Marx casing for high voltage isolation, and the gas mixture of the N_2 and SF_6 with the pressure of 90 kPa was sent to the switch case. In single mode, a maximum 540 kV high-voltage pulse was delivered to the water load with the rise-time of 28 ns and the pulse width of 95 ns, as shown in Figure 15. The peak power of the output pulse was calculated to be about 5 GW ($P = U^2/R$) and the E/V and P/W of the PFN-Marx generator body were about 6.5 J/L and 90 MW/kg, respectively. The generator was also tested in burst mode. The load voltage pulse-train of the generator on the water load, which comprises five pulses at a repetition rate of 30 Hz, is shown in Figure 16. The charging voltage of the PFN-Marx is 45 kV and the amplitude of the load voltage is about 500 kV, which was measured via a resistance voltage divider. The dispersion of the load voltage is less than 3%. The load voltage waveform indicates that the operation of the generator in single mode agrees well with that in burst mode.

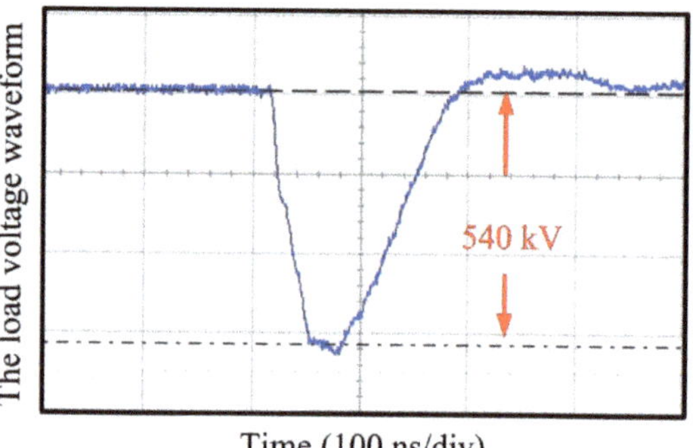

Figure 15. The load voltage waveforms on the water load.

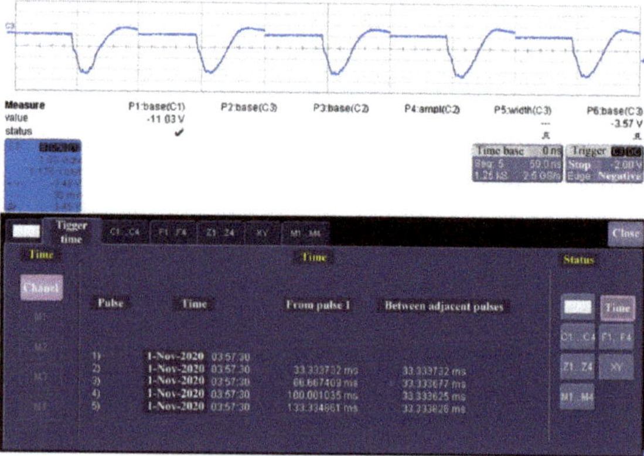

Figure 16. Screen shot of 5 pulses at 30 Hz operation.

However, the rectangular characteristic of the load voltage waveform in the experiment was not as good as in the simulation. The reason for that contains five parts. Firstly, the generator is constructed by multiple PFN modules, which are like building blocks. So, the coupling capacitor (C_c) between the adjacent PFN modules inevitably exists between the PFN modules, and the C_c, together with the PFN node inductors, forms a parasitic transmission line (PTL) with a short characteristic time. The final performance of the effect of the PTL is that the high-frequency oscillation is superimposed on the load voltage waveform. This is the main reason why the load voltage waveform's quality in the experiment was lower than that in the simulation. Secondly, the capacitor in the simulation is an ideal model and the mica in the practical experiment has self-inductance. In further work, the measure method of the mica capacitor's self-inductance will be studied. Thirdly, the inductance of the water resistor also affects the quality of the load waveform. Fourth, the load voltage was measured by a water resister divider with a low impedance type. The sample resistance of the resistor divider is 1 Ω. The deterioration of the rise-edge of the load voltage waveform is significantly affected by the length of the connecting line of the resistance divider, which is the inductance of the connecting line (L_m). The dying oscillation of the load voltage waveform is decreased as the L_m decreases. Finally, in the simulation analysis, the capacitance of the switch and the ground capacitance are ignored. These parasitic parameters also affect the load pulse. Actually, an oscillation damping circuit, which is to add an R-C unit between the last two PFN modules, has been preliminarily investigated in an experiment in another PFN-Marx system [17]. So, future work will also focus on the improvement of the load waveform's quality.

4.2. Experimental Results on a TTO

A miniaturized transit time oscillator (TTO) driven by this generator was initially investigated in an experiment. The setup for TTO operations is similar to the water load testing, with the dummy load removed. The Marx casing was filled with 160 kPa of pure SF_6, and, depending on the charging voltage, 80 kPa gas of N_2 mixed with 15% SF_6 was sent to the enclosed switch column. The generator was operated in triggering mode, and the output voltage and output current of the generator were measured by a resistance divider and a hand-made Rogovski coil, respectively. The typical waveforms of the generator and the microwave are shown in Figure 17. The pulse characteristics of the voltage waveform, including the voltage of 450 kV, the rise-time of 25 ns, and the pulse width of 90 ns, were measured in the oscilloscope. Obviously, the pulse width of the current waveform is shorter than that of the voltage waveform. The reason could be that the Rogovski coil has the defect during hand making. The sampling loop was not controlled to a minimum and the sampling resistor is not an inductive resistor. Besides, because the Rogovski coil is compact, there may be breakdown inside it during the measuring process. Therefore, distortion of the current signal exists. During the experiment, the Rogovski coil was used only for monitoring the electron beam production. A microwave pulse with power of about 100 MW and a pulse duration of 25 ns is generated when the diode voltage is 450 kV. The HPM signal was estimated to have a peak of 100 MW. The initial experimental results show the ability of the generator to drive the HPM source.

The TTO driven by the PFN-Marx generator in burst mode was also investigated initially in an experiment. The experimental results show that the TTO at 30 Hz with five pulses works well. The microwave power is about 85 MW, with the diode voltage about 440 kV, as shown in Figure 18. Obviously, there are some differences in the consistency of the microwave pulses. The experimental test with the PFN-Marx generator driving the TTO is still under study and detailed results will be published in the next work.

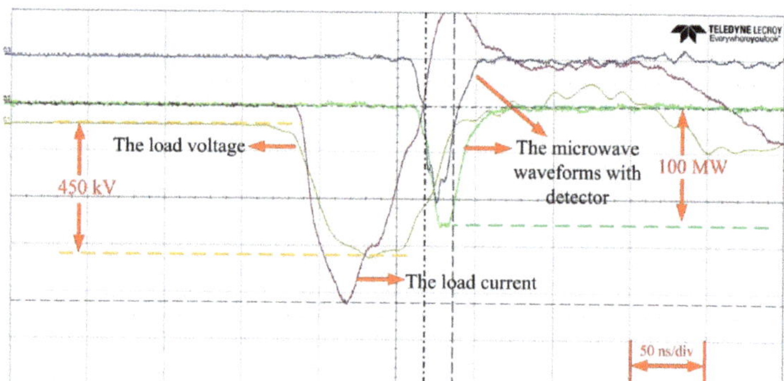

Figure 17. The typical waveforms on the transit time oscillator (TTO).

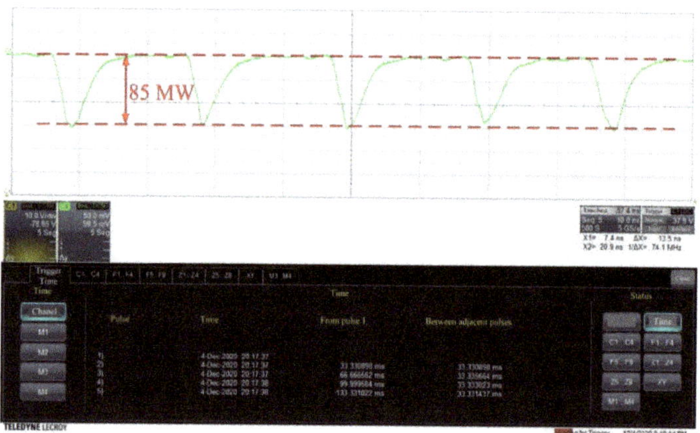

Figure 18. Screen shot of 5 microwave pulses at 30 Hz operation.

5. Summary

The 22-stage compact PFN-Marx generator based on the mica capacitors has shown successful 5 GW output on a water load. The performance of this PFN-Marx generator in burst mode is also described and the load voltage waveform indicates that the operation of the generator in single mode agrees well with that in burst mode. Typical pulse characteristics include a pulse width of 95 ns and a rise-time of 28 ns. A TTO HPM source driven by this generator was also tested and a 100 MW, 25 ns microwave pulse was obtained in an experiment.

A series of methods for compact and modular design of the PFN-Marx generator were used, including the large energy density mica capacitors, a mini-Marx trigger source with the integration of the magnetic transformer and switch, the PFN module design, a common switch case, and the connection between the PFN module and the switch. Under the condition of these designs, a single PFN module can be taken out from the PFN-Marx individually. Finally, the length of the PFN-Marx generator body is limited to 80 cm and its weight is about 56 kg. The ratio of the energy storage to volume and the ratio of power to weight of the 22-stage PFN-Marx generator are up to 6.5 J/L and 90 MW/kg, respectively.

Actually, a lot of work could be done to improve the performance of the generator; for example, increasing the pulse number and improving the quality of the output waveform, as well as optimizing the operation of the TTO source driven by the generator. Furthermore, based on our generator, other types of HPM sources, including a relativistic magnetron and a relativistic backward wave oscillator, are being studied.

Author Contributions: PFN-Marx design and structure design, H.Z. (Haoran Zhang); validation, T.S.; primary energy subsystem, S.L.; methodology, Z.Z.; TTO design, L.S.; control subsystem, H.Z. (Heng Zhang); writing—original draft preparation, H.Z. (Haoran Zhang); writing—review and editing, T.S. and Z.Z. All authors have read and agreed to the published version of the manuscript.

Funding: This research received no external funding.

Institutional Review Board Statement: Not applicable.

Informed Consent Statement: Not applicable.

Data Availability Statement: The data that support the findings of this study are available from the corresponding author upon reasonable request.

Acknowledgments: The authors wish to thank Bo Liang and Jia Li for their support during the assembly and the experiment.

Conflicts of Interest: The authors declare no conflict of interest.

References

1. Hansjoachim, B. Pulsed Power Systems/Principles and Applications. In *Pulsed Power Systems: Principles and Applications*; Springer: Berlin/Heidelberg, Germany, 2006.
2. Korovin, S.D.; Gubanov, V.P.; Gunin, A.V. Repetitive nanosecond high-voltage generator based on spiral forming line. In Proceedings of the Pulsed Power Plasma Science, Las Vegas, NV, USA, 17-22 June 2001.
3. Yang, J.; Zhang, Z.; Yang, H. Compact intense electron-beam accelerators based on high energy density liquid pulse forming lines. *Matter Radiat. Extrem.* **2018**, *3*, 278–292. [CrossRef]
4. McDonald, K.F.; Slenes, K. Compact modulator for high power microwave systems. In Proceedings of the Conference Record of the Twenty-seventh International Power Modulator Symposium, Arlington, VA, USA, 14–18 May 2006.
5. Spahn, E.; Buderer, G.; Wenning, W. A compact pulse forming network, based on semiconducting switches, for electric gun applications. *IEEE Trans. Magn.* **1999**, *35*, 378–382. [CrossRef]
6. Walter, J.; Dickens, J.; Kristiansen, M. Performance of a compact triode vircator and Marx generator system. In Proceedings of the IEEE Pulsed Power Conference, Washington, DC, USA, 28 June–2 July 2009.
7. Zhang, H.; Li, Z.; Zhang, Z. Investigation on the generation of high voltage quasi-square pulses with a specific two-node PFN-Marx circuit. *Rev. Sci. Instrum.* **2020**, *91*, 024702. [CrossRef] [PubMed]
8. Glasoe, G.N.; Lebacqz, J.V. *Pulse Generators*; McGraw-Hill: New York, NY, USA, 1948.
9. Song, F.; Zhang, B.; Li, C. Development and testing of a three-section pulse-forming network and its application to Marx circuit. *Laser Part. Beams* **2019**, *37*, 1–7. [CrossRef]
10. Rainer, B.; Jean-Pierre, D.; Sylvain, P. Modular, ultra-compact Marx generators for repetitive high-power microwave systems. *J. Korean Phys. Soc.* **2011**, *59*, 3522.
11. Neuber, A.A.; Chen, Y.J.; Dickens, J.C.; Kristiansen, M. A compact, repetitive, 500kV, 500 J, Marx generator. In Proceedings of the Pulsed Power Conference, Monterey, CA, USA, 13–15 June 2007.
12. Lassalle, F.; Morell, A.; Loyen, A.; Chanconie, T. Development and test of a 400-kV PFN Marx with compactness and rise time optimization. *IEEE Trans. Plasma Sci.* **2018**, *46*, 3313–3319. [CrossRef]
13. Nunnally, C.; Lara, M.; Mayes, J.R.; Nunnally, W.C. A compact 700-kV erected pulse forming network for HPM applications. In Proceedings of the 2011 IEEE Pulsed Power Conference, Chicago, IL, USA, 19–23 June 2011; pp. 1372–1376.
14. Mayes, J.R.; Mayes, M.G.; Lara, M.B. A novel Marx generator topology design for low source impedance. In Proceedings of the Pulsed Power Conference, Monterey, CA, USA, 13–15 June 2005.
15. Nunnally, C.; Mayes, J.R.; Hatfield, C.W.; Downden, J.D. Design and performance of an ultra-compact 1.8-KJ, 600-KV pulsed power system. In Proceedings of the IEEE Pulsed Power Conference, Washington, DC, USA, 28 June–2 July 2009.
16. Zhang, Y.; Liu, J. A new kind of solid-state Marx generator based on transformer type magnetic switches. *Laser Part. Beams* **2013**, *31*, 239–248. [CrossRef]
17. Li, Z.; Yang, J.; Hong, Z. Impact of parasitic transmission line on output of PFN-Marx generator. *High Power Laser Part. Beams* **2016**, *04*, 28.

 *electronics*

MDPI

Article

Novel Dual Beam Cascaded Schemes for 346 GHz Harmonic-Enhanced TWTs

Ruifeng Zhang [1], Qi Wang [1], Difu Deng [1], Yao Dong [1], Fei Xiao [1], Gil Travish [2] and Huarong Gong [1,*]

[1] School of Electronic Science and Engineering, University of Electronic Science and Technology of China, Chengdu 610054, China; rfzhang@std.uestc.edu.cn (R.Z.); wq1017697898@163.com (Q.W.); uestcddf@163.com (D.D.); dong.yaoyao@hotmail.com (Y.D.); fxiao@uestc.edu.cn (F.X.)

[2] ViBo Health LLC, Westwood Plaza, Los Angeles, CA 90095, USA; gtravish@gmail.com

* Correspondence: hrgong@uestc.edu.cn

Abstract: The applications of terahertz (THz) devices in communication, imaging, and plasma diagnostic are limited by the lack of high-power, miniature, and low-cost THz sources. To develop high-power THz source, the high-harmonic traveling wave tube (HHTWT) is introduced, which is based on the theory that electron beam modulated by electromagnetic (EM) waves can generate high harmonic signals. The principal analysis and simulation results prove that amplifying high harmonic signal is a promising method to realize high-power THz source. For further improvement of power and bandwidth, two novel dual-beam schemes for high-power 346 GHz TWTs are proposed. The first TWT is comprised of two cascaded slow wave structures (SWSs), among which one SWS can generate a THz signal by importing a millimeter-wave signal and the other one can amplify THz signal of interest. The simulation results show that the output power exceeds 400 mW from 340 GHz to 348 GHz when the input power is 200 mW from 85 GHz to 87 GHz. The peak power of 1100 mW is predicted at 346 GHz. The second TWT is implemented by connecting a pre-amplification section to the input port of the HHTWT. The power of 600 mW is achieved from 338 GHz to 350 GHz. The 3-dB bandwidth is 16.5 GHz. In brief, two novel schemes have advantages in peak power and bandwidth, respectively. These two dual-beam integrated schemes, constituted respectively by two TWTs, also feature rugged structure, reliable performance, and low costs, and can be considered as promising high-power THz sources.

Keywords: terahertz; traveling-wave tube; folded waveguide (FWG); slow wave system; high harmonic traveling wave tube

Citation: Zhang, R.; Wang, Q.; Deng, D.; Dong, Y.; Xiao, F.; Travish, G.; Gong, H. Novel Dual Beam Cascaded Schemes for 346 GHz Harmonic-Enhanced TWTs. *Electronics* **2021**, *10*, 195. https://doi.org/10.3390/electronics10020195

Received: 25 December 2020
Accepted: 14 January 2021
Published: 16 January 2021

Publisher's Note: MDPI stays neutral with regard to jurisdictional claims in published maps and institutional affiliations.

1. Introduction

Terahertz (THz) devices are widely used in high data-rate communication systems, plasma diagnostic, hazardous material detection, medical imaging, etc. However, the development of THz technology faces some challenges, such as lack of THz sources with high power, miniaturization, and low cost. Semiconductor THz sources and vacuum electronic THz sources are two common ones. Although semiconductor THz sources can produce high-power output up to milliwatt, they are usually troubled by over-high upfront cost. As a compromise, vacuum electronic devices (VEDs) may deliver higher output power with lower cost [1–7]. In 2004, a kind of compact THz free electron laser device was introduced by Stuart R A, with 1 kW pulsed power from 0.3 THz to 3 THz [8]. In 2010, Khanh Nguyen developed a high-gain multi-beam traveling wave tube (TWT) whose operating frequency range varies from 200 GHz to 250 GHz [9]. In 2011, Istok proposed a series of Backward-wave oscillators (BWOs) in which the grating line is utilized as slow wave structure (SWS). These devices can deliver several milliwatts output power at 1.4 THz [10]. In 2012, Paoloni et al. presented a cascade backward-wave amplifier operating at 1 THz [11]. From 2012 to 2016, Tucek et al. discussed a series of vacuum electronic amplifiers including one 100 mW 670 GHz prototype device driven by a novel solid-state source [12], a compact, microfabricated vacuum electronic amplifier with 39.4 mW of maximum output power from 0.835 THz to 0.875 THz [13], and a 29 mW 1.03 THz vacuum electronic with 20 dB of saturated gain and 5 GHz of instantaneous bandwidth [14]. In 2018, a folded waveguide (FWG) TWT is fabricated by Armstrong et al., with over 300 mW power in 231.5–235 GHz [15]. In 2020, Pan Pan et al. proposed a G-band continuous wave TWT. Saturated power of 20 W is generated from 217 GHz to 219.4 GHz [16].

For the development of nuclear fusion energy, the understanding of a critical plasma phenomenon as transport of the plasma is necessary. The collective Thompson scattering at THz frequency has been proven to be an adequate technology to map anomalous density fluctuation of electrons in the plasma, without perturbing its plasma behavior. The optically pumped far-infrared (FIR) laser is a practical radiation source for this technology and applied in the National Spherical Torus Experiment (NSTX) [17]. However, the features of the high price, large volume, and relatively low power restrict its mapping region. To extend the dimension of the plasma diagnostic, the BWOs operating at 346 GHz can be promising devices due to low cost, large output power, easy assembly, and compact volume. C. Paoloni et al. designed a 0.4 W double corrugated waveguide (DCW) BWO and a 1 W double-staggered grating (DSG) BWO [18]. J. Feng et al. designed a Grooved Single Grating (GSG) structure for 346 GHz BWO. The GSG circuit was fabricated by UV-lithographie galvanoformung abformung (LIGA) microelectromechanical technologies [19].

It should be noted that the BWO is strict about power source because BWO requires very suitable power source to maintain frequency stable. Phase noise of THz signal generation, which caused by power supply voltage ripple, should be reduced to ensure low bit-error rate in THz communication [20]. To reduce requirement of high-performance power source and obtain pure frequency spectrum of output power, we developed a kind of THz source, named high-harmonic TWT (HHTWT) [6,21]. Based on HHTWT, one new HHTWT and two novel types of THz sources operating at 346 GHz are proposed in this paper to improve the output power. The HHTWT can generate the THz band electromagnetic (EM) wave by amplifying the E-band signal. Compared with conventional THz signal sources, the application of high-power E-band signal source in HHTWT can input considerable signal into SWS. It can avoid input signal being interfered with and even drown out noise, which is caused by electron gun and discordance of SWS fabrication.

One of the novel THz sources, named cascaded enhanced HHTWT (CE-HHTWT), outputs the THz power by amplifying the signal, which is generated by HHTWT. The other one, named Pre-amplified HHTWT (PA-HHTWT), amplifies the THz band EM wave by inputting a relatively high power of fundamental wave into HHTWT.

This paper is organized as follows: HHTWT and two novel types of THz sources are introduced in Section 2, Section 3, and Section 4. Each section contains the operating principle, SWS design methodology, and simulation results. The analysis and design works are accomplished by CST particle studio. Section 5 is a summary of this paper.

2. HHTWT

2.1. Operating Principle of HHTWT

A HHTWT operating at 346 GHz is introduced firstly, which utilizes FWG as SWS. FWG is a promising type of SWS with wide bandwidth and high power. Compared with other conventional SWSs such as helix, FWG is of easy fabrication and assembling. Within it, the wave transmission path can be folded back upon itself multiple times, with a beam tunnel passing through its center. Energy exchange is achieved by synchronizing both the longitudinal energy flow speed and the electron beam velocity.

The schematic of the HHTWT SWS is shown in Figure 1. Port 1 and 4 are input and output ports, respectively. Attenuators are applied to match two severed ports (Port 2 and 3). The electron beam is sent from the electron gun into the tunnel. The SWS consists of three sections: Modulation section, drift tube, and radiation section. The modulation section operates at E band. The radiation section operates at THz band, corresponding to the fourth harmonic of input signal. There is also a drift tube between the modulation section and radiation section, where the EM wave is cutoff and only electron beam can pass.

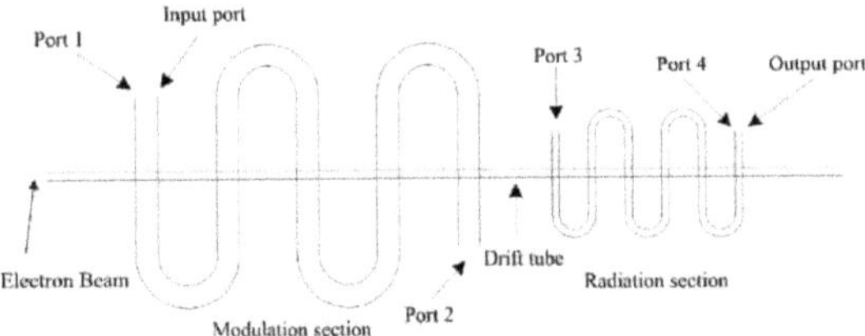

Figure 1. The slow wave structure (SWS) of high-harmonic traveling wave tube (HHTWT).

In the modulation section, the velocity of electron beam is modulated by the E-band input signal. When the electron beam traverses the drift tube, the velocity modulation is transformed into longitudinal density modulation. If the cutoff frequency of the radiation section is 300 GHz, the fundamental wave and lower-order harmonic waves can be cut off in the radiation section, and then only high-frequency EM waves are excited and amplified by the high-order harmonic beam current. Hence, when inputting a signal at 86.5 GHz, we can get a 346 GHz output signal.

Compared with conventional FWG TWT, HHTWT is featured by adopting two high frequency structures operating in two different bands. The power of conventional THz band FWG TWT is restricted due to the high loss. It leads to low gain and overlong SWS. Generally, the low input power of THz source could exacerbate the deleterious effects. The introduction of modulation section could modulate the electron beam efficiently. It is also instructive to mitigate the demand for a THz high-power signal source, by using a millimeter-wave high-power source.

2.2. SWS Design Methodology of HHTWT

2.2.1. SWS Design

Two sections of the HHTWT, i.e., the modulation section and the radiation section, have different operating bands. The modulation section works at E band, and the radiation section operates at THz band. The dispersion curve and interaction impedance are obtained by 3D EM simulation. For the modulation section, the phase shift at the center frequency is set to 1.41π, which ensures enough bandwidth. Then, interaction impedance should be set as high as possible. For the radiation section, the operating voltage should be the same with that in modulation section. The dispersion curve should also be flat to broaden bandwidth. Hence, the phase shift at the center frequency is set to 1.49π. Figure 2 shows the dispersion curve, associated electron beam line, and interaction impedance of the modulation and the radiation sections, respectively. The structural parameters of the HHTWT are shown in Table 1, in which a, b, h, p, and r are the width of the broad edge of the waveguide, the length of the narrow side of the waveguide, the height of the straight rectangular waveguide, the axial period length, and the radius of the electron beam channel, respectively. Between the modulation section and the radiation section, the drift tube is adopted to decrease the risk of self-oscillation and realize the transition from velocity modulation to density modulation. The length determination of two sections and drift tube are discussed later.

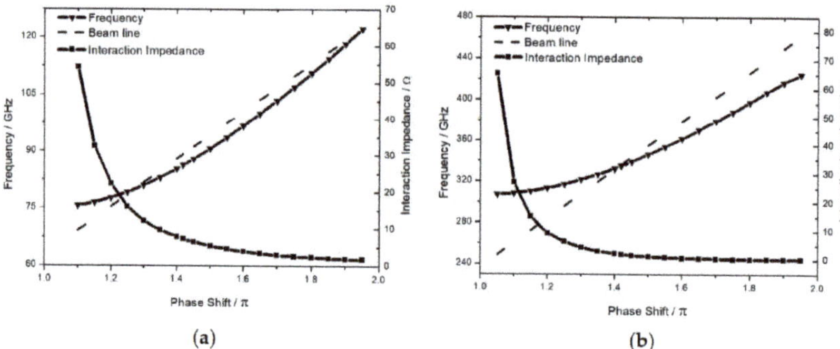

Figure 2. High frequency characteristics of the modulation and the radiation section. (**a**) The modulation section; (**b**) the radiation section.

Table 1. Structural parameters of the HHTWT.

Parameters	Modulation Section	Radiation Section	Drift Tube
a	2 mm	0.5 mm	/
b	0.32 mm	0.085 mm	/
p	0.64 mm	0.17 mm	/
h	0.51 mm	0.2 mm	/
r	0.08 mm	0.08 mm	0.08 mm

2.2.2. The SWS Length Determination

The following simulation results are obtained by CST Particle Studio. The operating voltage and current are set to 18.4 kV and 10 mA in Particle in Cell (PIC) simulation, respectively. With frequency increased, transmission loss caused by skin effect becomes significantly high. The surface roughness, determined by different manufacture technology, also has significant effect on the transmission loss [21,22]. The precision computer numerical control (CNC) lathe and electric discharge machining (EDM) are main and mature processing methods and widely applied in the fabrication of THz band and E-band SWS [22,23]. By referencing experimental cases in [22,23] and summarizing our engineering experience in [21], the effective conductivities of sections operating at E band and THz band are set as 3.5×10^7 S/m and 1×10^7 S/m, respectively.

Due to low input power, the length of the modulation section should be long enough to ensure deep electron beam modulation. Hence, we construct 100 periods of modulation SWS, and determine the length of modulation section depending on the bunch state of electron beam. The frequency of input signal is 86.5 GHz, and the input power is 200 mW.

The phase space of beam electron in the modulation section is shown in Figure 3. It shows the amplification process is at linear state before 38 mm. Hence, the length of modulation section is set as 38 mm.

In order to determine the length of the drift tube, a series of current monitors are placed every other millimeter on the drift tube. Figure 4 plots the Fourier transform of the electron current signal at the end of the drift tube. Fundamental component and other three high harmonic components are also shown in it. However, only the fourth harmonic wave can be excited and amplified because the fundamental wave and other lower-order harmonic waves are cut off in the radiation section. In general, the length of the drift tube is controlled at the position where the fourth harmonic current is the largest. Figure 5 depicts the relative amplitude of the fourth harmonic current curve versus the length of the drift tube. The length of the drift tube is determined as 12 mm. Figure 6 shows the phase space graph of the electron beam at the end of the drift tube. It depicts that the electron beam is modulated by EM wave and stay in the linear state. The energy is still stored in the electron beam.

The length of the radiation section is chosen where output power reaches saturation point. Figure 7 shows the power versus the length of the HHTWT at 346 GHz. Hence, the length of the radiation section is 34 mm. The phase space graph of electron beam at the end of radiation section is shown as Figure 8.

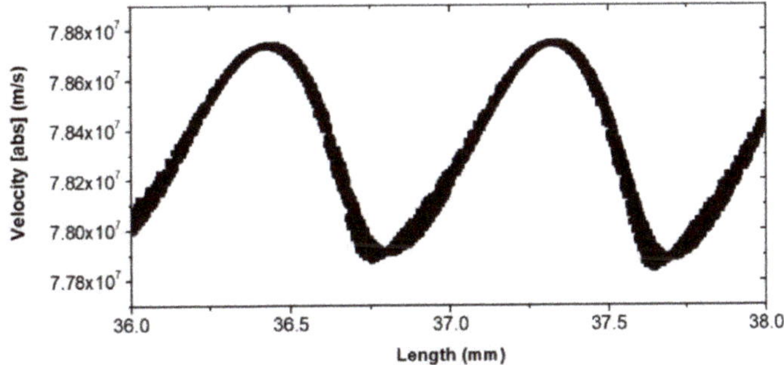

Figure 3. The phase space graph of the electron beam in the modulation section.

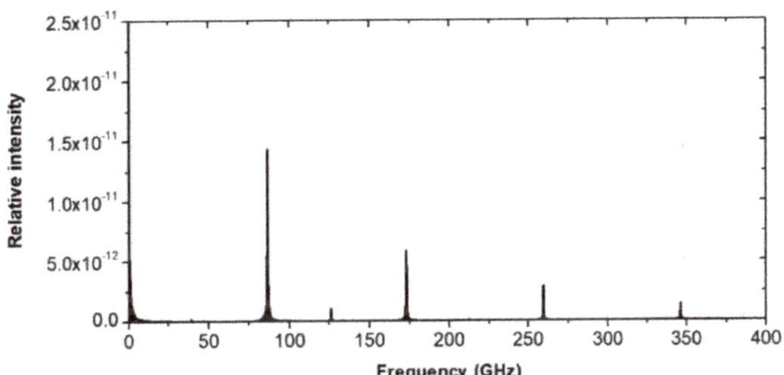

Figure 4. The Fourier transform of the electronic current signal.

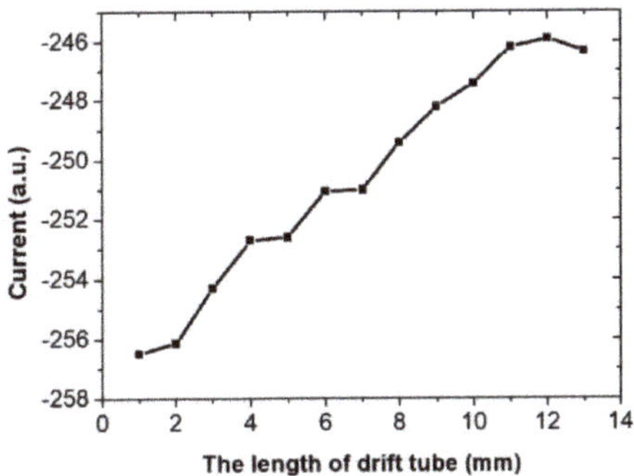

Figure 5. The graph of amplitude of the fourth harmonic.

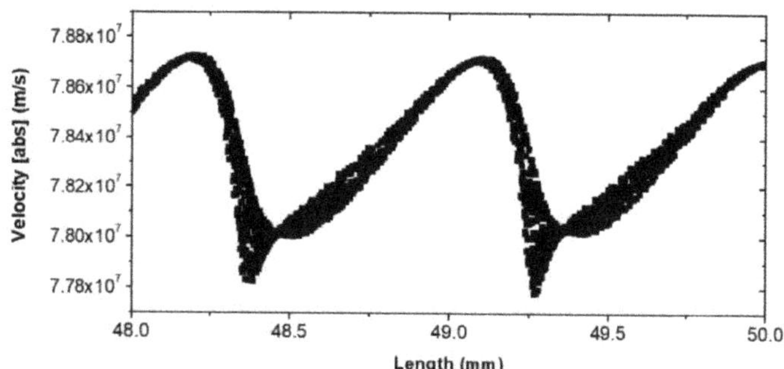

Figure 6. The phase space graph of electron beam at the end of drift tube.

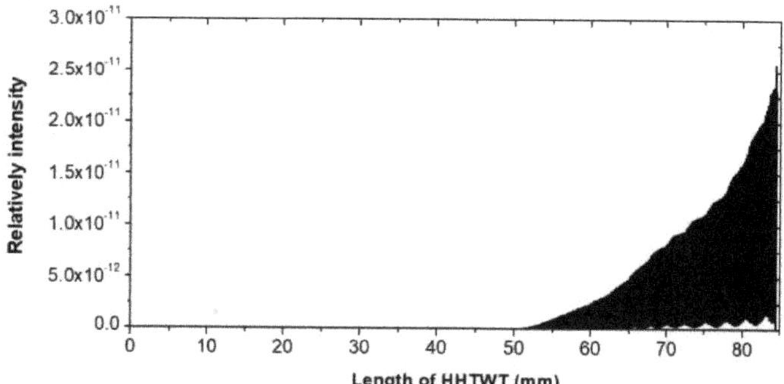

Figure 7. Power curve of the HHTWT at 346 GHz.

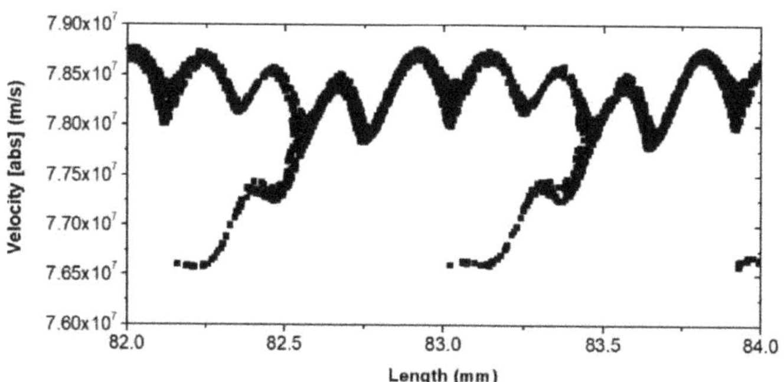

Figure 8. The phase space graph of the electron beam at the end of the radiation section.

Figure 8 shows the fourth harmonic is excited. Compared with Figure 6, the velocity of central cluster decreased significantly in Figure 8. It means the electron beam transfers a lot of energy to the EM wave, during the beam wave interaction in the radiation section. For convenience of reference, the parameters of HHTWT are summarized in Table 2.

Table 2. Parameters of the HHTWT.

Parameters	Values
Beam voltage	18.4 kV
Beam current	10 mA
Beam radius	0.04 mm
Length of modulation section	38 mm
Length of drift tube	12 mm
Length of radiation section	34 mm

2.3. The Simulation Results

The output power of the HHTWT is shown in Figure 9. It can deliver about 300 mW at 346 GHz. As illustrated in Figure 10, the spectrum of the output power is concentrated at 346 GHz. Figure 11 plots the output power from 340 to 348 GHz. When 200-mW signal is input from 85 to 87 GHz, over 100-mW power can be delivered with the 8-GHz bandwidth.

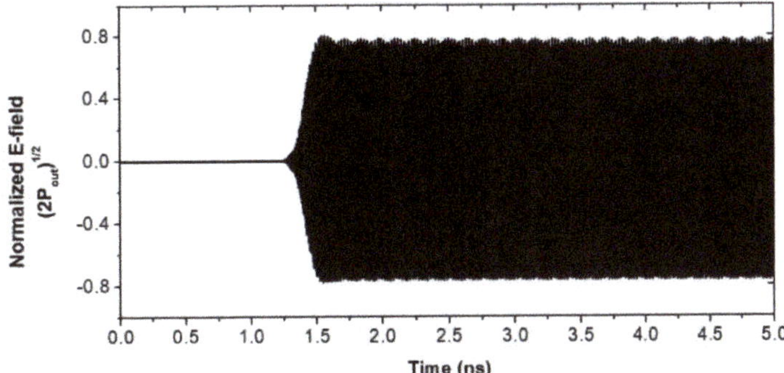

Figure 9. The output signal of the HHTWT at 346 GHz.

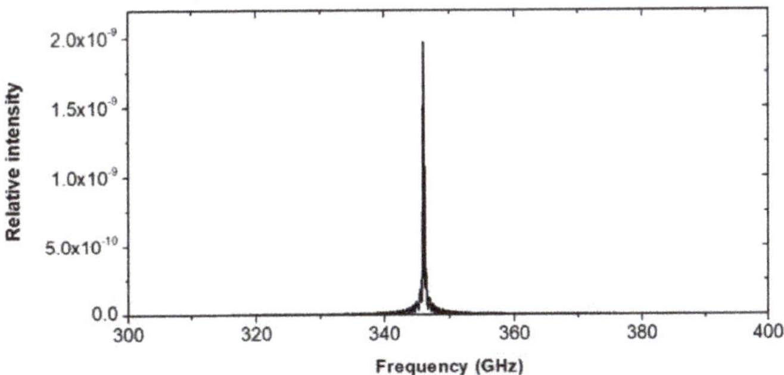

Figure 10. The Fourier transform of output signal at 346 GHz.

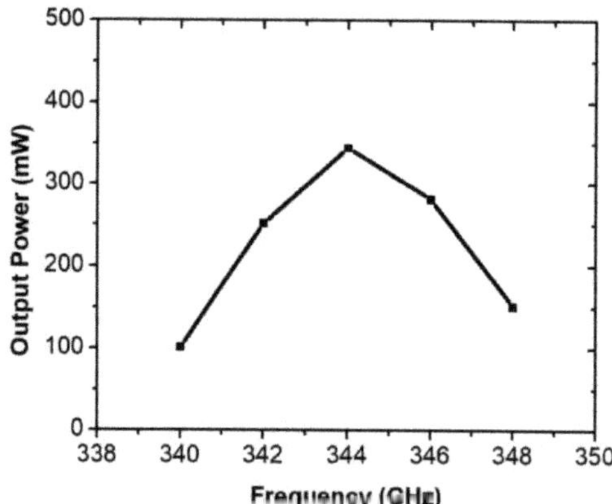

Figure 11. Output power of HHTWT.

3. Cascaded Enhanced HHTWT

3.1. Operating Principle of CE-HHTWT

In order to amplify the power of HHTWT, a novel structure is proposed on the basis of HHTWT, named cascaded enhanced HHTWT. It is featured by introduction of another electron beam that forms dual-beam THz band CE-HHTWT.

CE-HHTWT is demonstrated in Figure 12. Connecting an amplification section to the output port of the HHTWT forms dual-beam THz band CE-HHTWT. The THz signal enters the amplification section from the radiation section. Beam-wave interaction occurs between a new electron beam and input signal of amplification section. Compared with the HHTWT, the efficiency and the output power of dual-beam CE-HHTWT is improved.

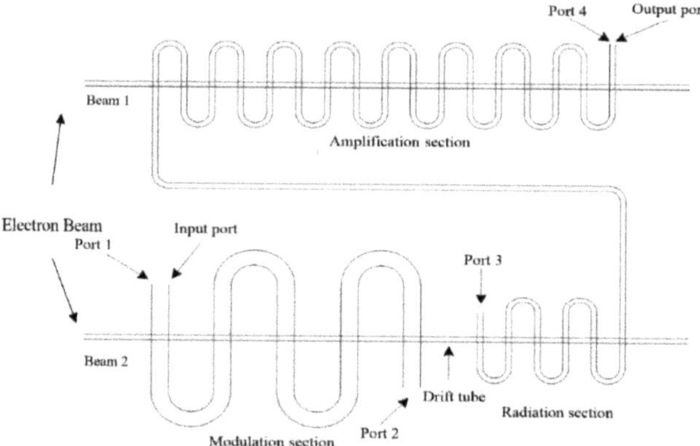

Figure 12. The SWS of the dual-beam CE-HHTWT.

3.2. SWS Design Methodology of CE-HHTWT

As shown in Figure 12, the amplification section operates at THz band. To simplify design and fabrication, the amplification utilizes the radiation section with the same structural parameters as SWS does. The parameters of two electron beams in CE-HHTWT are the same with those in HHTWT.

The length of the amplification section is chosen where the output power reaches saturation point. Figure 13 shows the power curve versus the length of the amplification section at 346 GHz. Hence, the length of the amplification section is 34 mm. The parameters of CE-HHTWT are summarized in Table 3.

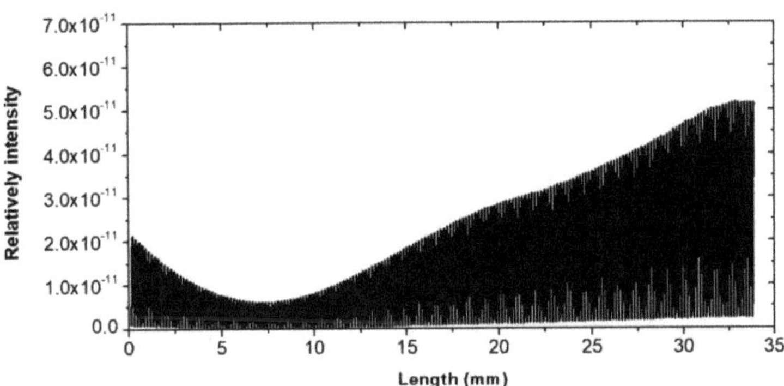

Figure 13. Power curve of the amplification section.

Table 3. Parameters of the CE-HHTWT.

Parameters	Values
Beam voltage of beam 1 and 2	18.4 kV
Beam current of beam 1 and 2	10 mA
Beam radius	0.04 mm
Length of modulation section	38 mm
Length of drift tube	12 mm
Length of radiation section	34 mm
Length of amplification section	34 mm

3.3. The Simulation Results

The output power of the CE-HHTWT is shown in Figure 14. The input signal is 200 mW at 86.5 GHz. Furthermore, a signal of 1100 mW is obtained at 346 GHz.

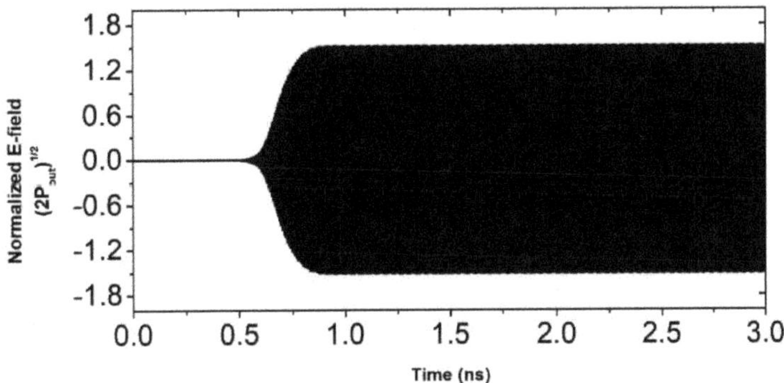

Figure 14. The output signal of the CE-HHTWT at 346 GHz.

Figure 15 is the Fourier transform of signal at end of the amplification section. The spectrum of the output signal is concentrated at the frequency of 346 GHz.

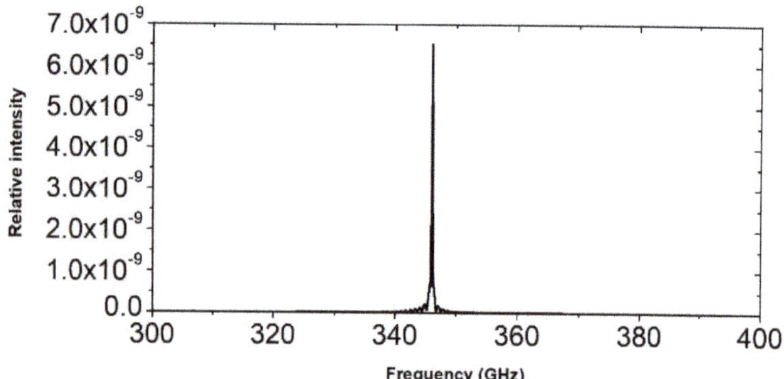

Figure 15. The Fourier transform of output signal.

Figure 16 shows the gain property and bandwidth property of CE-HHTWT. At 346 GHz, the output power is saturated when the input power is 200 mW. Figure 16b shows that more than 400 mW output power could be achieved from 340 to 348 GHz. The maximum power is 1100 mW at 346 GHz. The 3-dB bandwidth of CE-HHTWT is 5 GHz.

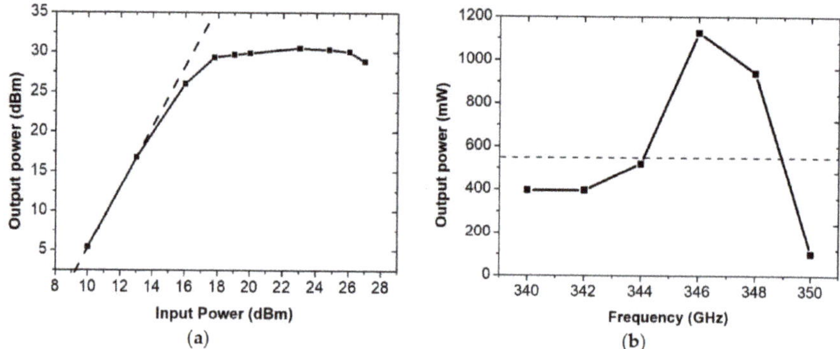

Figure 16. Gain property (**a**) and bandwidth property (**b**) of CE-HHTWT.

4. Pre-Amplified HHTWT

4.1. Operating Principle of PA-HHTWT

The PA-HHTWT is presented in Figure 17. A pre-amplification section is connected to the input port of a HHTWT, which constitute the other new TWT. Fundamental signal is amplified by an electron beam firstly. Then, as an input signal, the amplified fundamental signal interacts with a new electron beam to generate the THz signal in the HHTWT.

In comparison with HHTWT and PA-HHTWT, the electron beam is deeply modulated. Hence, it can improve the output power.

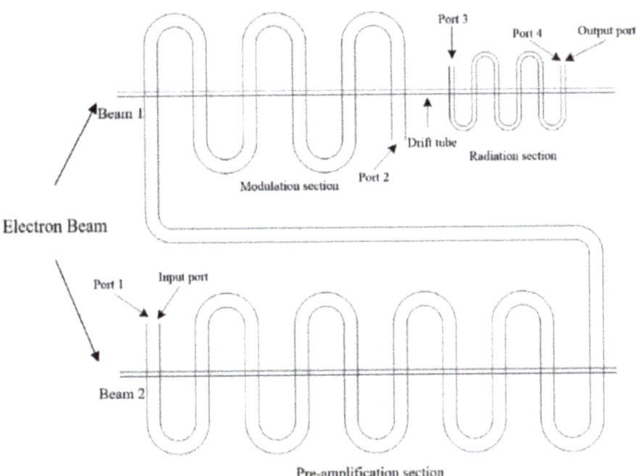

Figure 17. The SWS of dual beam PA-HHTWT.

4.2. SWS Design Methodology of PA-HHTWT

For the PA-HHTWT, the pre-amplification section plays a role in amplifying the fundamental wave to several-watts level. The pre-amplification section utilizes the same structural parameters with modulation section. The length of pre-amplification section is set as 51 mm. The power versus axial length curve and the output signal are shown in Figures 18 and 19, respectively. The input power is set at 200 mW and then 4.5-W output power is obtained.

The length of the modulation section is chosen to make the energy modulate the beam as deeply as possible, rather than amplify the EM wave. Figure 20 depicts the input and output signal of the modulation section at 86.5 GHz. When the length of the modulation section is 20.48 mm, the output power is barely higher than the input power, because the energy of electron beam is almost not changed within the modulation process. The behavior mentioned above is also validated in Figure 21. The electron beam is modulated more deeply than HHTWT and CE-HHTWT. Meanwhile, there is no obvious nonlinear characteristic.

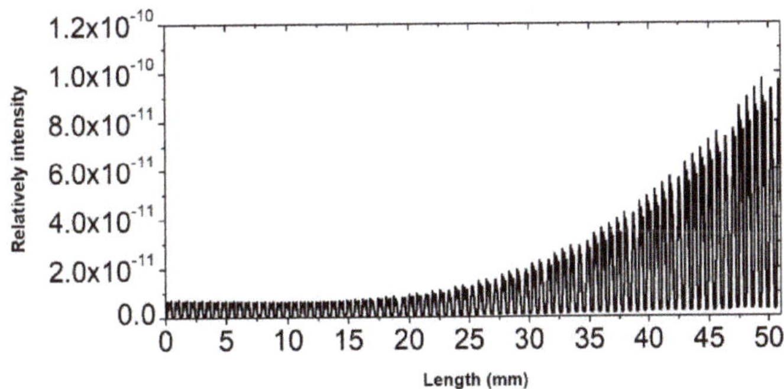

Figure 18. Power curve of the pre-amplification section.

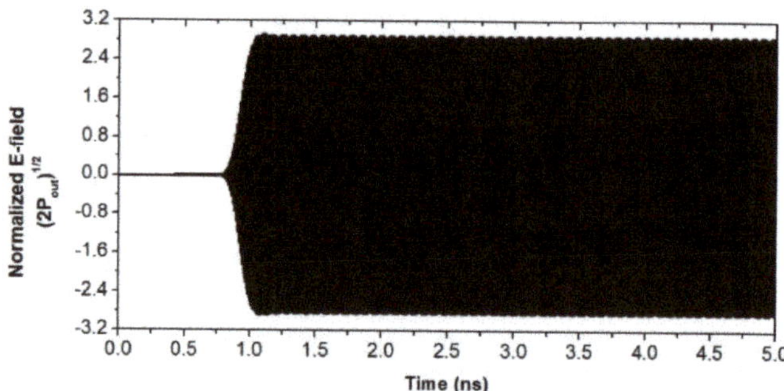

Figure 19. Output signal of pre-amplification section.

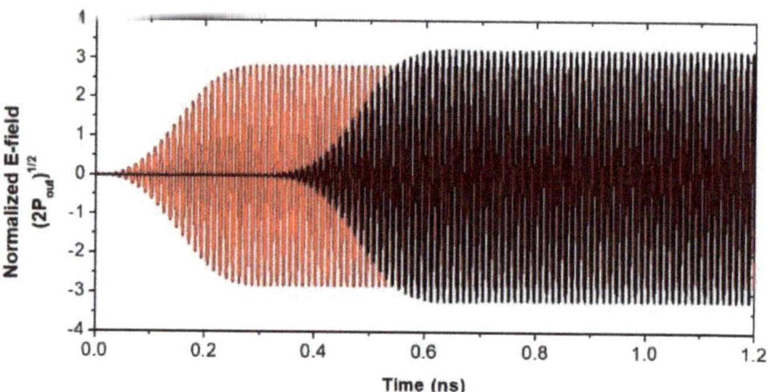

Figure 20. Input (red) and output (black) power of the modulation section at 86.5 GHz.

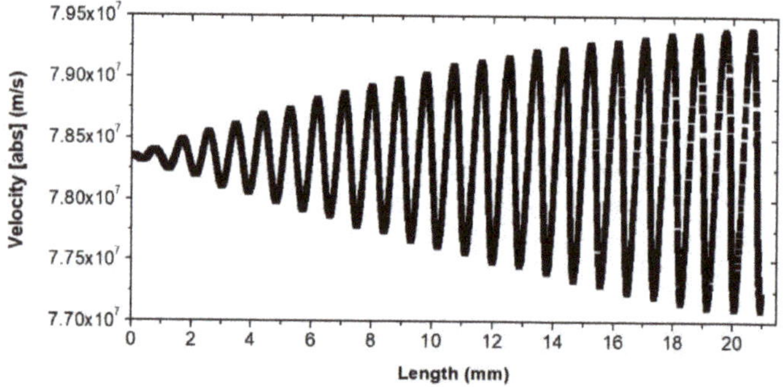

Figure 21. The phase space graph of electron beam in modulation section of PA-HHTWT.

The method to determine the length of the drift tube and the radiation section is similar to that in the HHTWT and CE-HHTWT. Eventually, the length of the drift tube and the radiation section are chosen at 4 mm and 10.2 mm, respectively. For convenience, the parameters of PA-HHTWT are shown in Table 4.

Table 4. Parameters of the PA-HHTWT.

Parameters	Values
Beam voltage of beam 1 and 2	18.4 kV
Beam current of beam 1 and 2	10 mA
Beam radius	0.04 mm
Length of pre-amplification section	51 mm
Length of modulation section	20.5 mm
Length of drift tube	4 mm
Length of radiation section	10.2 mm

4.3. The Simulation Results

The output power of PA-HHTWT at 346 GHz is shown in Figure 22. The output power is 970 mW. The spectrum of output signal is shown in Figure 23.

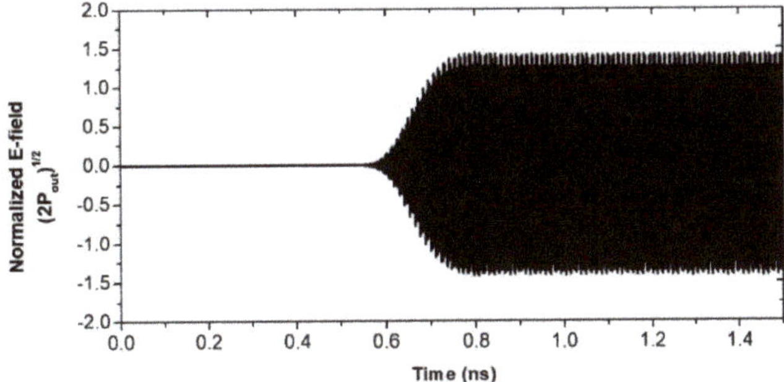

Figure 22. The output signal of the PA-HHTWT at 346 GHz.

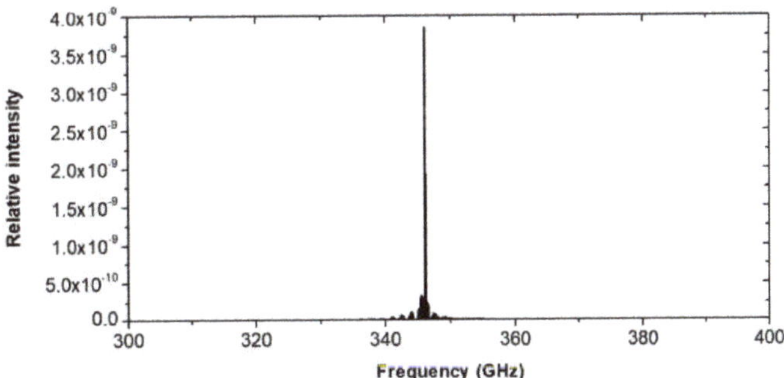

Figure 23. The Fourier transform of output signal.

Figure 24a demonstrates the gain property of PA-HHTWT. The 200-mW input power can drive the device to saturated state. The output power of PA-HHTWT is over 600 mW in 338–350 GHz with 200-mW driven power from 84.5 GHz to 87.5 GHz, as shown in Figure 24b. The 3-dB bandwidth of PA-HHTWT is 16.5 GHz.

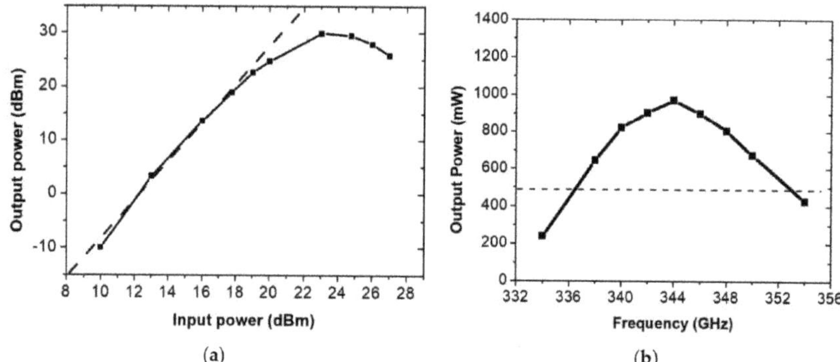

Figure 24. Gain property (**a**) and bandwidth property (**b**) of PA-HHTWT.

5. Conclusions

To develop high-power THz sources, HHTWT operating at 346 GHz is introduced and analyzed in this paper. It can output THz signal by amplifying the fourth harmonic component of E-band RF signal. The simulation results demonstrate that amplifying the fourth harmonic signal is a promising way to obtain high-power THz signal. Furthermore, two power-enhanced schemes, CE-HHTWT and PA-HHTWT, are proposed in this paper. CE-HHTWT yields the THz power by amplifying the signal generated by HHTWT. PA-HHTWT amplifies THz band EM wave by importing high power-level fundamental wave into HHTWT. The operating principle and design methodology of two schemes are also described in this paper, including high frequency characteristics and length determination of each sections. To validate working principle which relies upon Pierce's linear theory, the simulation results are predicted by CST. Driven by two 18.4 kV,10 mW electron beams, CE-HHTWT can generate over 400 mW power in 340–348 GHz by inputting signal of 200 mW from 85 GHz to 87 GHz. The peak power is 1100 mW at 346 GHz. The output power of PA-HHTWT is over 600 mW in 338–350 GHz. The 3 dB bandwidth reaches 16.5 GHz. The simulation results show that CE-HHTWT and PA-HHTWT have advantages in peak power and wide bandwidth, respectively. Compared with conventional THz sources driven by high-power solid-state power amplifier, these two schemes, both constituted by two TWTs, have superiorities in structural strength and costs.

In future study, CE-HHTWT and PA-HHTWT SWSs can be fabricated by using high-precision CNC milling. Periodic permanent magnet (PPM) and dual-beam electron gun will be adopted in electron optical system. To validate the feasibility of design, the insertion loss, output power, bandwidth, and phase noise characters will be tested. Thus, the schemes of CE-HHTWT and PA-HHTWT are hopeful approaches to realize practical high-power THz sources for plasma diagnostic, communication, and radar.

Author Contributions: Conceptualization, R.Z., H.G., and Q.W.; methodology, R.Z. and H.G.; software, R.Z. and D.D.; validation, R.Z. and Y.D.; formal analysis, R.Z. and H.G.; investigation, R.Z., G.T., and H.G.; writing—original draft preparation, R.Z.; writing—review and editing, R.Z. and F.X. All authors have read and agreed to the published version of the manuscript.

Funding: This work was supported in part by the National Natural Science Foundation of China under Grant 61370152, 61531010 and in part by the American Physical Society's International Travel Grant Award Program.

Conflicts of Interest: The authors declare no conflict of interest.

References

1. Booske, J. Vacuum electronic sources for high power terahertz-regime radiation. In Proceedings of the 2011 IEEE International Vacuum Electronics Conference (IVEC), Bangalore, India, 21–24 February 2011; p. 11. [CrossRef]
2. Bates, D.J.; Ginzton, E.L. A traveling-wave frequency multiplier. *Proc. IRE* **1957**, *45*, 938–944. [CrossRef]
3. Liu, S.; Zhong, R. Recent development of terahertz science and its applications. *J. Univ. Electron. Sci. Technol. China* **2009**, *38*, 481–486.
4. Wang, W. *Microwave Engineering Technology*, 2nd ed.; National Defense Industry Press: Beijing, China, 2009.
5. Karetnikova, T.A.; Ryskin, N.M.; Belov, K.V.; Ploskih, A.E. Simulation of a 0.2 thz second-harmonic multiplier with sheet electron beam. In Proceedings of the 2017 10th UK-Europe-China Workshop on Millimetre Waves and Terahertz Technologies (UCMMT), Liverpool, UK, 11–13 September 2017; pp. 1–2. [CrossRef]
6. Gong, H.; Travish, G.; Xu, J.; Wei, Y.; Feng, J.; Gong, Y. High-power tunable terahertz radiation by high-order harmonic generation. *IEEE Trans. Electron Devices* **2013**, *60*, 482–486. [CrossRef]
7. Zhang, C. Research on High-Power Millimeter-Wave Folded Waveguide Traveling Wave Tube. Ph.D. Thesis, University of Electronic Science and Technology of China, Chengdu, China, 2011.
8. Stuart, R.A.; Wright, C.C.; Al-shamma'a, A.I. Compact tuneable terahertz source. In Proceedings of the 2006 European Microwave Conference, Manchester, UK, 10–15 September 2006; pp. 322–324. [CrossRef]
9. Nguyen, K.; Ludeking, L.; Pasour, J.; Pershing, D.; Wright, E.; Abe, D.K.; Levush, B. Design of a high-gain wideband high-power 220-GHz multiple-beam serpentine TWT. In Proceedings of the 2010 IEEE International Vacuum Electronics Conference, Monterey, CA, USA, 18–20 May 2010; pp. 23–24. [CrossRef]
10. Galdetskiy, A.V.; Golenitskiy, I.I.; Myakinkov, V.U.; Negirev, A.A.; Rudy, U.B. On power consumption reduction in 700 GHz BWO. In Proceedings of the 2011 IEEE International Vacuum Electronics Conference, Bangalore, India, 21–24 February 2011; pp. 57–58. [CrossRef]
11. Paoloni, C.; Mineo, M.; Di Carlo, A.; Durand, A.J.; Krozer, V.; Kotiranta, M. 1-thz cascade backward wave amplifier. In Proceedings of the IVEC 2012, Monterey, CA, USA, 24–26 April 2012; pp. 237–238. [CrossRef]
12. Tucek, J.C.; Basten, M.A.; Gallagher, D.A.; Kreischer, K.E.; Lai, R.; Radisic, V.; Leong, K.; Mihailovich, R.A. 100 mw, 0.670 thz power module. In Proceedings of the IVEC 2012, Monterey, CA, USA, 24–26 April 2012; pp. 31–32. [CrossRef]
13. Tucek, J.C.; Basten, M.A.; Gallagher, D.A.; Kreischer, K.E. 0.850 thz vacuum electronic power amplifier. In Proceedings of the IEEE International Vacuum Electronics Conference, Monterey, CA, USA, 22–24 April 2014; pp. 153–154. [CrossRef]
14. Tucek, J.C.; Basten, M.A.; Gallagher, D.A.; Kreischer, K.E. Operation of a compact 1.03 thz power amplifier. In Proceedings of the IEEE International Vacuum Electronics Conference (IVEC), Monterey, CA, USA, 19–21 April 2016; pp. 1–2. [CrossRef]
15. Armstrong, C.M.; Kowalczyk, R.; Zubyk, A.; Berg, K.; Meadows, C.; Chan, D.; Schoemehl, T.; Duggal, R.; Hinch, N.; True, R.B.; et al. A Compact Extremely High Frequency MPM Power Amplifier. *IEEE Trans. Electron Devices* **2018**, *65*, 2183–2188. [CrossRef]
16. Pan, P.; Tang, Y.; Bian, X.; Zhang, L.; Lu, Q.; Li, Y.; Feng, Y.; Feng, J. A G-Band Traveling Wave Tube With 20 W Continuous Wave Output Power. *IEEE Electron Device Lett.* **2020**, *41*, 1833–1836. [CrossRef]
17. Lee, K.C.; Domier, C.W.; Johnson, M.; Luhmann, N.C.; Park, H. FIR laser tangential interferometry/polarimetry on NSTX. *IEEE Trans. Plasma Sci.* **2004**, *32*, 1721–1726. [CrossRef]
18. Paoloni, C.; Gamzina, D.; Himes, L.; Popovic, B.; Barchfeld, R.; Yue, L.; Zheng, Y.; Tang, X.; Tang, Y.; Pan, P.; et al. THz Backward-Wave Oscillators for Plasma Diagnostic in Nuclear Fusion. *IEEE Trans. Plasma Sci.* **2016**, *44*, 369–376. [CrossRef]
19. Feng, J.; Tang, Y.; Gamzina, D.; Li, X.; Popovic, B.; Gonzalez, M.; Himes, L.; Barchfeld, R.; Li, H.; Pan, P.; et al. Fabrication of a 0.346-THz BWO for Plasma Diagnostics. *IEEE Trans. Electron Devices* **2018**, *65*, 2156–2163. [CrossRef]
20. Krupnov, A.F. Phase Lock-In of MM/SUBMM Backward Wave Oscillators: Development, Evolution, and Applications. *Int. J. Infrared Millim. Waves* **2001**, *22*, 1–18. [CrossRef]
21. Gong, H.; Wang, Q.; Deng, D.; Dong, Y.; Xu, J.; Tang, T.; Su, X.; Wang, Z.; Gong, Y.; Travish, G. Third-harmonic traveling-wave tube multiplier-amplifier. *IEEE Trans. Electron Devices* **2018**, *65*, 2189–2194. [CrossRef]
22. Gamzina, D.; Li, H.; Himes, L.; Barchfeld, R.; Popovic, B.; Pan, P.; Letizia, R.; Mineo, M.; Feng, J.; Paoloni, C.; et al. Nanoscale Surface Roughness Effects on THz Vacuum Electron Device Performance. *IEEE Trans. Nanotechnol.* **2016**, *15*, 85–93. [CrossRef]
23. Pan, P.; Li, H.; Feng, J. Study on loss of folded waveguide structures of 220 GHz and 340 GHz. In Proceedings of the 2016 IEEE International Vacuum Electronics Conference (IVEC), Monterey, CA, USA, 19–21 April 2016; pp. 1–2. [CrossRef]

Article

Investigation on 220 GHz Taper Cascaded Over-Mode Circular Waveguide TE_{0n} Mode Converter

Tongbin Yang [1], Xiaotong Guan [2,3], Wenjie Fu [1,3,*], Dun Lu [1], Xuesong Yuan [1,3] and Yang Yan [1,3]

1   School of Electronic Science and Engineering, University of Electronic Science and Technology of China, Chengdu 610054, China; yangtongbin@std.uestc.edu.cn (T.Y.); ludun@std.uestc.edu.cn (D.L.); yuanxs@uestc.edu.cn (X.Y.); yanyang@uestc.edu.cn (Y.Y.)
2   School of Physics, University of Electronic Science and Technology of China, Chengdu 610054, China; guanxt@uestc.edu.cn
3   Terahertz Science and Technology Key Laboratory of Sichuan Province, University of Electronic Science and Technology of China, Chengdu 610054, China
*   Correspondence: fuwenjie@uestc.edu.cn

Abstract: This paper proposes a taper cascaded over-mode circular waveguide TE_{0n} mode converter for the millimeter and terahertz wave gyrotron. The mode converter of this structure can effectively reduce the difficulty of high frequency mode converter in fabrication. This paper verifies the feasibility of this new structure from theory, simulation, and experiment. Based on coupled wave theory calculations, three TE_{02}-TE_{01} mode converters with lengths of 65.43 mm (4 segments), 119.3 mm (6 segments) and 136 mm (8 segments) and a TE_{03}-TE_{02} mode converter with a length of 92 mm (8 segments) are optimized. The conversion efficiency in the frequency band 215–225 GHz is 91.8–94%, 93–95%, 95–98.78% and 95–98.44%. Because the length of the mode converter is clearly limited, this paper selects the TE_{02}-TE_{01} mode converter with a length of 65.43 mm (4 segments) and the TE_{03}-TE_{02} mode converter with 92 mm (8 segments) for simulation and experimental verification. In the simulation software Computer simulation technology (CST), the TE_{02}-TE_{01} and TE_{03}-TE_{02} mode converters and their composed TE_{03}-TE_{01} mode converters are selected for modeling and analyzing. The simulation results and theoretical calculation results of the three mode converters only have different degrees of frequency deviation, and the frequency deviation of the 4-stage TE_{02}-TE_{01} mode converter can be ignored; the frequency deviations of TE_{03}-TE_{02} mode converter and TE_{03}-TE_{01} mode converter are 2 GHz and 3 GHz, respectively. The experimental system is a field scanning system based on a vector network analyzer (VNA), which scans the input and output of the mode converter respectively. The experimental result is that when the input mode purity is 92% in TE_{01} mode, the output mode TE_{03} mode has a mode purity of 82%, and it has lower transmission loss. In this paper, the results from theory, simulation and experiment are in good agreement. This type of mode converter is easy to prepare, which makes it an effective alternative for high frequency curvilinear waveguide mode converter.

Keywords: mode converter; 220 GHz; taper

Citation: Yang, T.; Guan, X.; Fu, W.; Lu, D.; Yuan, X.; Yan, Y. Investigation on 220 GHz Taper Cascaded Over-Mode Circular Waveguide TE_{0n} Mode Converter. *Electronics* **2021**, *10*, 103. https://doi.org/10.3390/electronics10020103

Received: 23 December 2020
Accepted: 5 January 2021
Published: 6 January 2021

Publisher's Note: MDPI stays neutral with regard to jurisdictional claims in published maps and institutional affiliations.

1. Introduction

Vacuum electronic devices (VEDs) (e.g., Traveling-Wave Tube (TWT), Backward-Wave Oscillator (BWO), Magnetron, Klystron, Gyrotron) are important high-power microwave radiation sources for scientific, industrial and military applications [1]. As the VEDs operating frequency extends to the millimeter wave and terahertz wave bands, the size of the interaction circuit shrinks, and the power capability is consequently reduced. Thus, operating at high-order mode is proposed and attempted. Gyrotron, as a fast wave VED, is successfully operated at high-order modes. Thumm et al. reported that the gyrotron operated at the highest $TE_{32,19}$ (Transverse Electric) mode [2]. However, the high-order waveguide modes are symmetrical nonlinear polarization modes, and the axial radiation is an unsatisfactory hollow state, which is not conducive to long-distance transmission and direct application under the condition of over-mode operation. Depending on the working conditions, these modes need to be converted. There are two types of mode converters for VEDs. One is a waveguide mode converter which is usually used for higher-order volume modes (TE_{0n}, $n > 1$) [3], and its advantages are high converter efficiency, low reflectivity, and no parasitic oscillation of the device; another is quasi-optical mode converter which is used for higher-order whispering gallery modes (TE_{mn}, $m > 1$, $n = 1, 2$) [4] and its advantages are high power capacity, low transmission loss, and high output mode purity.

For a waveguide mode converter, the mode conversion scheme is normally as follows:

(1) TE_{0n}-TE_{01}-TE_{11}-HE_{11} [5,6]
(2) TE_{0n}-TE_{01}-TM_{11}-HE_{11} [7–9]

within which TE_{01} mode has low loss at high frequencies and is suitable for long-distance transmission; HE mode is a hybrided mode of TE (Transverse Electric) mode and TM (Transverse Magnetic) mode. HE_{11} includes TE_{11} (84% power) and TM_{11} (16% power). HE_{11} mode has the characteristics of linear polarization and Gaussian distribution, and is suitable for antenna transmission and application [6].

In both conversion sequences, the mode converting from TE_{0n} mode to TE_{01} mode is foremost. The TE_{0n}-TE_{01} waveguide mode converter changes its radial characteristics through radius perturbation. A common method is a corrugated circular waveguide mode converter with cyclically sinusoidal radius change [3]. At a low frequency, the sinusoidal radius change is easy to achieve. As the operating frequency increases, the sensitivity to processing errors increases and the difficulty of its implementation also increases exponentially. When the operating frequency increase up to terahertz wave region, the fabrication is extremely difficult and expensive.

In this paper, an economical gradual-radius cascaded circular waveguide mode converter is proposed as shown in Figure 1, whose fabrication is much easier than conventional type. A prototype of TE_{03}-TE_{02}-TE_{01} conversion for a frequency tunable 220 GHz gyrotron [10] is designed, manufactured, and tested. The results present a good performance at 215–225 GHz and would be an expectable approach in terahertz wave applications. This paper is organized as follows. The second part introduces the theoretical calculation and simulation results of the mode converter and gives the design ideas, main formulas, and structural parameters of the design. The third part introduces the experimental process and results analysis, and Section 4 draws the conclusion.

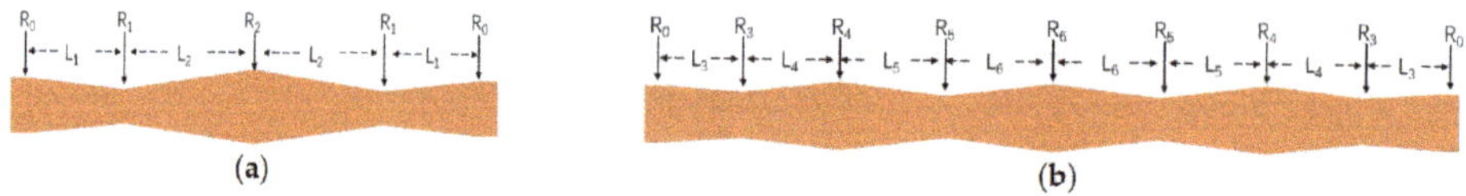

Figure 1. (a) Structure of TE_{02}-TE_{01} mode converter; (b) Structure of TE_{03}-TE_{02} mode converter.

2. Design Model and Simulation

2.1. Theoretical Calculation of Mode Converter

2.1.1. Theoretical Deduction

The inhomogeneous circular waveguide is divided into three categories: changes in transmission direction, changes in transmission medium, and changes in transmission radius. The mode converter uses the different transmission characteristics of each mode under the three changing conditions, selectively changes the three conditions, and obtains the required operating mode through mode-coupling. The common three types of circular waveguide mode converters are divided into serpentine waveguides with varying axis [11], dielectric-filled waveguides [12], and corrugated waveguides with varying transmission radius [13]. The theory of analyzing the coupling between modes in a waveguide is called coupled wave theory [14]. The design of the mode converter with radius perturbation is to solve the boundary value problem of coupled wave ordinary differential equations, the equations are as follows:

$$\frac{dA^+_{mn'}}{dz} = -\frac{1}{2}\frac{d(\ln\gamma_{mn'})}{dz}A_{mn'}{}^- - \gamma_{mn'}A^+_{mn'} + \sum_{+mn}A^+_{mn}C^+_{(mn')(mn)} + \sum_{-mn}A^-_{mn}C^-_{(mn')(mn)} \tag{1}$$

$$\frac{dA^-_{mn'}}{dz} = -\frac{1}{2}\frac{d(\ln\gamma_{mn'})}{dz}A_{mn'}{}^+ + \gamma_{mn'}A^-_{mn'} + \sum_{+mn}A^+_{mn}C^-_{(mn')(mn)} + \sum_{-mn}A^-_{mn}C^+_{(mn')(mn)} \tag{2}$$

where A^+_{mn}, A^-_{mn} is the amplitude of the mn mode wave and the superscript indicates the direction of propagation, γ_{mn} is the propagation constant of mn mode, and $\gamma_{mn} = \alpha_{mn} + j\beta_{mn}$, α_{mn} is the attenuation constant, β_{mn} is phase constant, both α_{mn} and β_{mn} are function of z, $C^\pm_{(mn')(mn)}$ is the coupling coefficient of mn mode and mn' mode in the same direction or reverse direction.

TE_{mn}-$TE_{mn'}$

$$C^\pm_{(mn')(mn)} = \frac{m^2\left(R_{mn'}X^2_{mn} \pm R_{mn}X^2_{mn'}\right) \mp \left(R_{mn} \pm R_{mn'}\right)X^2_{mn}X^2_{mn'}}{\left(R_{mn}R_{mn'}\right)^{1/2}\left(X^2_{mn} - m\right)^{1/2}\left(X^2_{mn'} - m\right)^{1/2}\left(X^2_{mn'} - X^2_{mn}\right)}\frac{1}{a}\frac{da}{dz}(-1)^{n+n'} \tag{3}$$

TE_{mn}-$TM_{mn'}$

$$C^\pm_{(mn')(mn)} = \frac{m}{\left(R_{mn}R_{mn'}\right)^{1/2}\left(X^2_{mn} - m\right)^{1/3}}\frac{1}{a}\frac{da}{dz}(-1)^{n+n'+1}, \tag{4}$$

Because the transmission direction and transmission medium of the mode converter with gradual radius have not changed and the coupling coefficient of the input mode TE_{0n} mode and TM mode is 0, only the coupling between TE modes can be considered in the calculation. The mode coupling of TE_{mn}-$TE_{mn'}$ satisfies $n' - n = 1$ and $m = 1$, so the coupling coefficient can be written as follows:

$$C^\pm_{(0n')(0n)} = \mp\frac{X_{0n}X_{0n'}\left(R_{0n} \pm R_{0n'}\right)}{\left(R_{0n}R_{0n'}\right)^{1/2}\left(X_{0n'} - X_{0n}\right)}\frac{1}{a}\frac{da}{dz}(-1)^{n+n'} \tag{5}$$

where X_{0n} ($X_{0n'}$) is the zero point value of Bessel function $J'_0(X_{0n})$ ($J'_0(X_{0n'})$),$R_{0n} = \beta_{0n}/k_0$ and $R_{0n'} = \beta_{0n'}/k_0$, k_0 is the free-space wavelength, a is the radius of the circular waveguide as a function of z.

Suppose the length of the mode converter is L, where $z = 0$ is the beginning of the circular waveguide and $z = L$ is the end of the circular waveguide. There is only one mode at the input port of the mode converter, the initial amplitude is 1. The output port of the waveguide does not have an aback-propagating wave. So the wave amplitude of each model at the input and output ports satisfy

$$A^+_{0n}|_{z=0} = [(1,0),(0,0),\cdots,(0,0)]^T \tag{6}$$

$$A^-_{0n}|_{z=L} = [(0,0),(0,0),\cdots,(0,0)]^T, \tag{7}$$

where the first element of the vector $\left[A^+_{0n}\right]$ is the amplitude of the input working mode, the second element is the amplitude of the output working mode, the other elements each represent a parasitic mode. Equations (6) and (7) and Equations (1) and (2) form the boundary value problem of the first-order nonlinear coupled wave differential equations. The forward wave amplitude A^+_{0n} and reverse wave amplitude A^-_{0n} along the z can be get to solve the equations.

The center working frequency of the Gyrotron is 220 GHz. The conversion of the output mode TE_{03} to TE_{01} needs to be realized in two stages, namely TE_{03}-TE_{02} and TE_{02}-TE_{01} mode conversion. Considering the power capacity of the mode converter and the output circular waveguide radius of the Gyrotron is 5 mm, the average radius of the mode converter's fluctuation is determined to be 5 mm. The main design indexes of the mode converter are the conversion efficiency, the working bandwidth, and the length of the converter. Editing the Matlab numerical calculation program uses the coupled wave theory and synthesizes three optimization design indicators. The optimization process is as follows: Given the initial number of cascaded segments N of the mode converter, a multi-parameter optimization model of the specific structural dimensions of each segment of the circular waveguide is established, and the main optimization goal is to maximize the conversion efficiency, taking into account the working bandwidth for optimization calculation. Change the number of cascades N and repeat the above optimization steps to obtain the most optimized design parameters. Compare the calculated results with the CST simulation results.

2.1.2. Calculation Results of the TE_{02}-TE_{01} Mode Converter

The structure of the 4-segment TE_{02}-TE_{01} mode converter is shown in Figure 1a, and the center frequency is 220 GHz. The numerical calculation takes into account factors such as multi-mode, reverse wave, and Ohmic loss, and uses the fourth-order Runge-Kutta method to perform optimization iterations, and the results of different segments are shown in Figure 2.

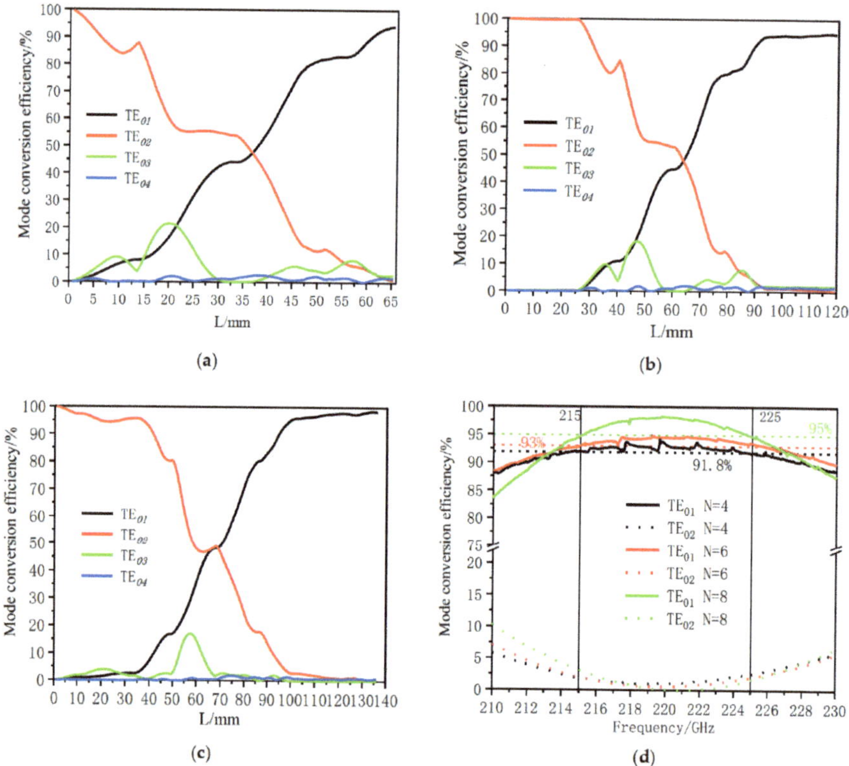

Figure 2. The calculation results of TE_{02}-TE_{01} mode converter: (**a**) Mode conversion efficiency along z of a 4-segment converter; (**b**) Mode conversion efficiency along z of a 6-segment converter; (**c**) Mode conversion efficiency along z of an 8-segment converter; (**d**) Mode conversion efficiency in the different frequencies.

In the process of the numerical, multi-mode, reverse wave, ohmic loss, and other factors are taken into account. The fourth-order and fifth-order Runge-Kutta methods were used to optimize iteratively, and the converters with different segment numbers were obtained. The optimization results are shown in Figure 2. From the Figure 2a–c, it can be concluded that the parasitic modes generated during the transmission process are TE_{03} and TE_{04} modes. The power excited in the first half of the converter is greater, and it is effectively suppressed in the second half. The length of the 4-stage cascaded mode converter is 65.43 mm, and the mode conversion efficiency is 94% at 220 GHz. The length of the 6-segment mode converter is 119.3 mm, and the mode conversion efficiency at 220 GHz is 94.78%. The length of the 8-segment mode converter is 136 mm, and the conversion efficiency at the center frequency point is 98.78%. The Figure 2d shown that the more segments of the cascade mode changer, the longer its length and the higher the conversion efficiency. In the working frequency band 215–225 GHz, the conversion efficiency of the three mode converters are 91.8–94%, 93–95%, 95–98.78%. Comparing the length and conversion efficiency of the converter, the length of the 6-segment is almost twice the length of the 4-segment, but the overall conversion efficiency is only increased by 1–2%, and the length of the 8-segment is only 16.7 mm longer than the length of the 6-segment, the conversion efficiency has increased by 3–4%.

2.1.3. Calculation Results of the TE_{03}-TE_{02} Mode Converter

The center frequency is 220 GHz, and the geometric structure of the TE_{03}-TE_{02} mode converter is shown in Figure 1b. Using the same numerical calculation method as TE_{02}-TE_{01} to design TE_{03}-TE_{02} mode converter, the final result of optimization is shown in Figure 3. It can be seen from the results that the parasitic modes of the converter are TE_{01} and TE_{04}, which are effectively suppressed during the mode conversion process. The length of the 8-segment TE_{03}-TE_{02} mode converter is 92 mm, and the mode conversion efficiency is 98.44% at 220 GHz. The conversion efficiency in the 215–224.2 GHz frequency band is higher than 95%.

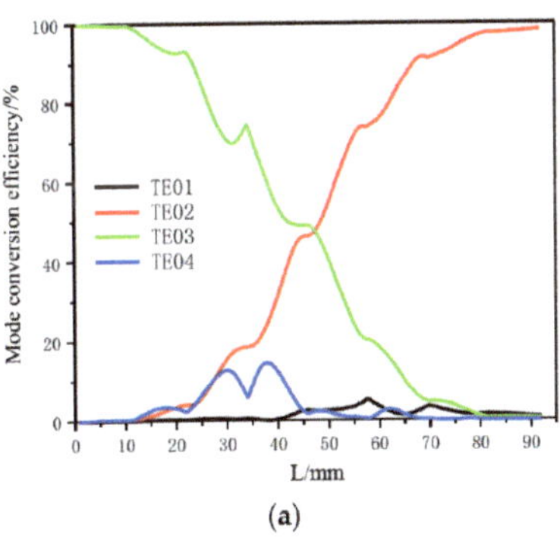
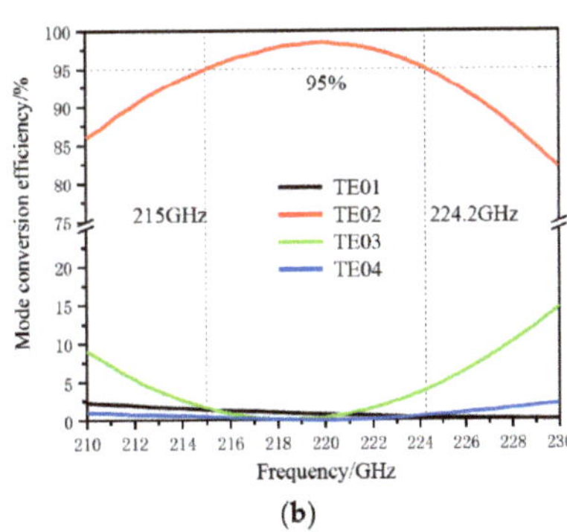

Figure 3. The calculation results of TE_{03}-TE_{02} mode converter: (**a**) Mode conversion efficiency along z of the mode converter; (**b**) Mode conversion efficiency in the different frequencies.

2.2. Simulation Results of Mode Converters

This mode converter is a part of the output structure of the gyrotron. Due to the limitation of experimental conditions, the total length of the two modes is required to be less than 200 mm. The design scheme of the processing experiment is to take four sections of TE_{02}-TE_{01} and six sections of TE_{03}-TE_{02} mode converters. The two mode converters are simulated and analyzed respectively, and the TE_{03}-TE_{01} mode converter composed of the two mode converters is analyzed and the calculation and simulation are also compared.

The mode generator in the experiment is a mode converter ($TE_{10}^{\square}$ to $TE_{01}°$) that converts from a rectangular waveguide TE_{10} mode to a round waveguide TE_{01} wave, and the mode converter is a reversible symmetrical two-port device. Therefore, the TE_{01} mode is used as the input mode in the comparison between simulation and calculation of the TE_{03}-TE_{01} mode converter. The structural dimensions of the calculated and simulated mode converters are shown in Tables 1 and 2.

Table 1. The structure size of TE_{02}-TE_{01} mode converter.

	R0	R1	R2	L1	L2
Calculation (mm)	5	3.79	6.14	13.9	19.3
Simulation (mm)	5	3.79	6.14	13.9	19.3

Table 2. The structure size of TE_{03}-TE_{02} mode converter.

	R3	R4	R5	R6	L3	L4	L5	L6
Calculation (mm)	4.87	5.32	4.5	5.54	9.3	11	12.2	11.8
Simulation (mm)	4.87	5.32	4.5	5.54	9.3	11	12.2	11.8

Perform simulation verification in the simulation software CST [15] with the dimensions in the table, and compare it with the calculated result as shown in Figure 4. The size of the mode converter calculated by the optimization is modeled separately in the three-dimensional simulation software CST. The simulation and theoretical calculation results of TE_{02}-TE_{01} and TE_{03}-TE_{02} mode converters are shown in Figure 4a,b. The S_{21} parameters of TE_{01}, TE_{02}, TE_{03} and TE_{04} in the simulation results are converted into the power content of the simulation output port, and compared with the power content of each mode output port calculated theoretically. The simulation results of the four-segment TE_{02}-TE_{01} mode converter are basically the same than the theoretical calculation results. The conversion efficiency in the working frequency band is above 91%. In the simulation results of TE_{03}-TE_{02} mode converter, the frequency band where the conversion efficiency of TE_{03} to TE_{02} remains above 95% is 217–227 GHz. Compared with the theoretical calculation result, there is a frequency offset of 2 GHz. The comparison results of other modes can also support this conclusion.

The content of each mode in the transmission direction of the TE_{01}-TE_{03} mode converter composed of two mode converters in the theoretical calculation at the operating frequency of 220 GHz is shown in Figure 4c. At the output port of the mode converter, the mode content of TE_{03} is 90%. In the simulation results, the change rule of the electric field transformation of the TE_{01}-TE_{03} mode converter in the transmission direction conforms to the trend of each mode transformation calculated in theory. In the working frequency band 215–225 GHz, the theoretically calculated conversion efficiency is 83.8–90%. The simulation result has a frequency deviation of 3 GHz compared with the calculation result. In the frequency band 218–230 GHz, the mode conversion efficiency is 81.2–90%. The results are shown in Figure 4d.

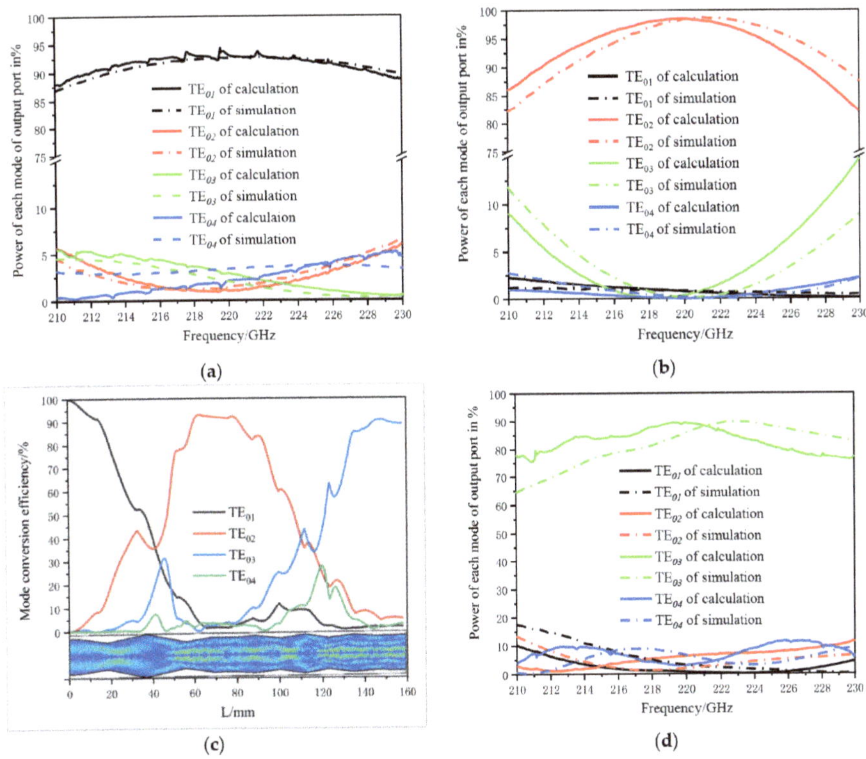

Figure 4. The calculation and simulation results of (**a**) 4-segment TE_{02}-TE_{01} mode conversion efficiency. (**b**) TE_{03}-TE_{02} mode conversion efficiency. (**c**) Calculation mode conversion efficiency and simulation E-Field of the TE_{03}-TE_{01} mode converter along the z direction. (**d**) TE_{03}-TE_{01} mode conversion efficiency.

3. Experimental Demonstration

The reversibility of the mode conversion can be used to measure the mode converter from the fundamental mode of the waveguide to the higher order mode. By measuring the S_{21} of the two groups of mode converters symmetrically connected, the conversion efficiency of the mode converter and the purity of the output port mode can be calculated. However, this mode converter measurement method is not suitable for the measurement of high frequency over-mode waveguide mode converters. The over-mode waveguide mode converter and the adapter from the standard waveguide to the over-mode waveguide connected at both ends will form a resonant cavity, making it easy to excite spurious modes and generate resonance. The S_{21} of the results cannot calculate the performance of the mode converter.

The test system in this paper is shown in Figure 5a. The 1-port spread spectrum module of the Vector Network Analyzer (VNA) and the device under test connected to it, are fixed on the optical platform. The 2-port spread spectrum module of the VNA and the rectangular waveguide probe connected to it are fixed on the 2D displacement platform (X-axis and Y-axis). One of the electric field components of the surface where the probe port is located can be tested through the 2D displacement platform scanning controlled by computer. The position of 1-port probe and the 2D platform remains unchanged and the 2-port probe is rotated 90 degrees along the Z-axis and then scanned through the 2D platform remains. Another electric field component of the surface can be tested. This kind of test scheme can measure the electric field component of the tested plane in the working frequency band of the vector network analyzer spread spectrum module. In the experiment of this article, first measure the two electric field components of the input of the mode converter ($E_{IN\text{-}X}$ and $E_{IN\text{-}Y}$).

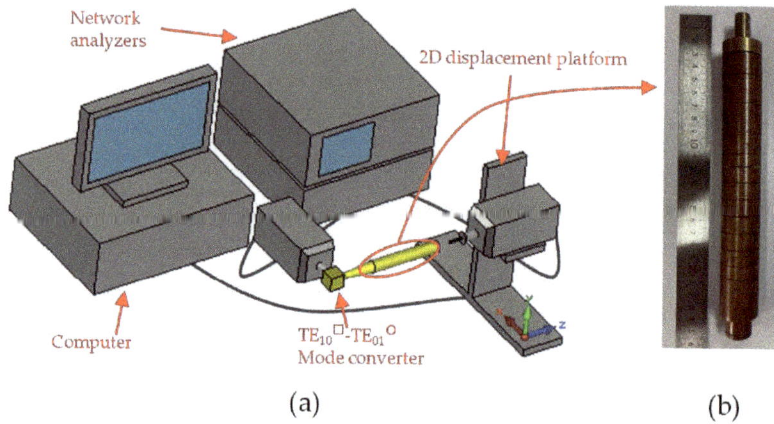

Figure 5. (**a**) The test systems of the converter and (**b**) The TE_{03}-TE_{01} mode converter after brazing.

The mode converter designed in this paper is composed of two mode converters TE_{02}-TE_{01} and TE_{03}-TE_{02}. These mode converters are segmented cascade structure. To ensure the complete connection of each component during the test, the two mode converters are integrally welded. The tested TE_{03}-TE_{01} mode converter is composed of TE_{03}-TE_{02} and TE_{02}-TE_{01} mode converters. The experimental system and the mode converter after welding are shown in Figure 5. The connecting parts between the mode converter and the VNA output waveguide are a $TE_{10}^{\square}$ to TE_{01}° mode converter and a circular waveguide adapter from 1.8 mm to 10 mm in diameter. The VNA in this system is made by the 41st Institute of China Electronics Technology Group Corporation (CETC).

Figure 6 is the experimental results of electric field component at input port of the mode converter. The test results of the electric field components in the X and Y directions at the 220 GHz input port are shown in Figure 6a,b. The blue dotted line in the figure is the ideal distribution of the corresponding electric field component to TE_{01} mode. It can be concluded from the comparison of the two field distribution that the two electric field components of the input port have a higher degree of matching with the electric field components of the standard TE_{01} mode.

Figure 7 is the experimental results of electric field component at output port of the mode converter. The electric field at the output port of the mode converter is tested in the X and Y directions. The results at the operating frequency of 220 GHz are shown in Figure 7a,b. It can be seen from the results in the figure that the mode of the output port is TE_{03} mode. Comparing the measured electric field with the ideal electric field distribution, it can be seen that the radius of the outmost peak ring of the TE_{03} mode is larger than the ideal value. The position of measuring the electric field distribution is a short distance away from the opening of the waveguide. The electric field distribution tested is the result of electromagnetic wave transmission after a certain distance in the air. Due to the large diffraction of TE_{03} mode in the air, the radius of the most ring is larger than the waveguide radius in the test results.

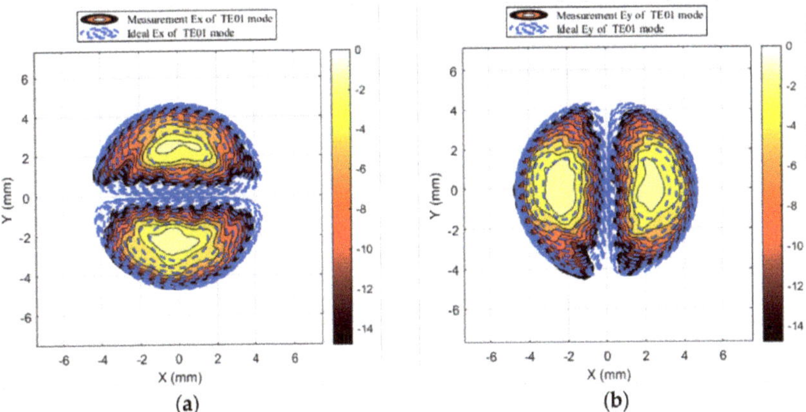

Figure 6. (**a**) The test results of E_x of TE_{01} mode; (**b**) the test results of E_y of TE_{01} mode.

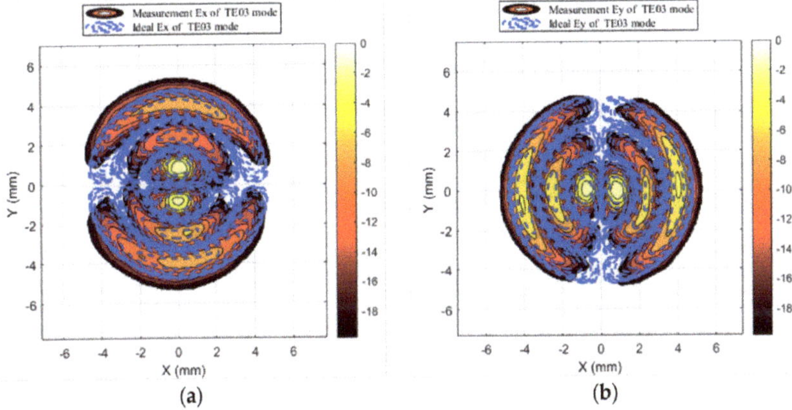

Figure 7. (**a**) The test results of E_x of TE_{03} mode; (**b**) the test results of E_Y of TE_{03} mode.

In order to study the performance of the tested mode converter more clearly, some frequency points of 215–225 GHz in the working bandwidth of the mode converter are selected for electric field scanning. Formula (8) can be used to calculate the mode purity of the input and output ports of the mode converter [16]. The calculation results of the purity of the electric field component measured by the input and output ports in the working frequency band is shown in Figure 8a.

$$\eta = \frac{\iint_s f(x,y) \bullet \psi^*(x,y)ds \bullet \iint_s f^*(x,y) \bullet \psi(x,y)ds}{\iint_s f(x,y) \bullet f^*(x,y)ds \bullet \iint_s \psi(x,y) \bullet \psi^*(x,y)ds} \tag{8}$$

where $f(x, y)$ represents the measured electric field distribution of the input and the output the mode converter, $\psi(x,y)$ is the ideal field distribution of the correspondence mode, the star (*) means the complex conjugation.

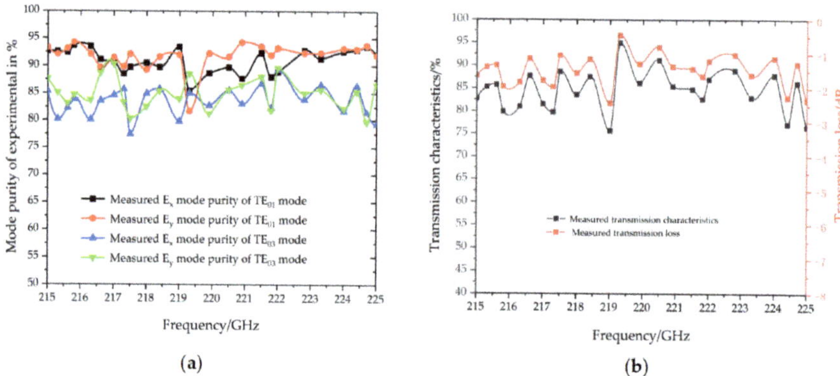

Figure 8. (**a**) The calculation result of the mode purity of the electric field component at the input and output of the mode converter, and (**b**) The experimental results of transmission loss of mode converter.

It can be seen from Figure 8a that at the input of the mode converter, the mode purity measured from the two electric field components is basically the same. The TE_{01} mode purity at the input port of the mode converter is maintained between 90–95% in the operating frequency band, except a small band nearly 215 GHz, the mode purity is reduced to 83%. The calculation results of the mode purity corresponding to the two electric field components of the TE_{03} mode measured at the output port are also basically the same. Mode purity of the TE_{03} remains at about 82% in the entire operating frequency band. The mode purity curves of the electric field components in the X direction and the Y direction have good consistency at the input and output ends, indicating that the scanning planes in the two directions are basically parallel, and the uniformity of the scanning field value is good, and it shows the experimental data is reliable.

Processing the scanned field value data can calculate the transmission loss of the mode converter. Integrate the measured two electric field components on the measured plane to get the power of the two electric field components (P_{IN-X} and P_{IN-Y}). Add the ingrate two components together, and express the sum as the input power of the mode converter as P_{IN} ($P_{IN} = P_{IN-X} + P_{IN-Y}$). The output power of the mode converter is measured as P_{OUT-X}, P_{OUT-Y} and P_{OUT} ($P_{OUT} = P_{OUT-X} + P_{OUT-Y}$) in same measured method. The S_{21} of the mode converter can be calculation by the calculation formula of transmission loss $S_{21} = P_{OUT}/P_{IN}$. The calculation results are shown in Figure 8b. The transmission characteristic of the mode converter is 82% in the operating frequency band and 75% in the vicinity of 219 GHz. At 219 GHz, the purity of the input mode is low, which increases the transmission loss of the millimeter wave signal.

In general, the results of the experiment show that the cascade mode converter outputs TE_{03} mode with 82% mode purity when the input mode is 92% pure TE_{01} mode. It is basically in line with the mode conversion efficiency of the theoretical and simulation results, and has lower transmission loss. It fully meets the requirements as a supporting device for gyrotron output.

4. Conclusions

In this paper, a taper cascaded over-mode circular waveguide TE_{03}-TE_{01} mode converter for a 220 GHz gyrotron has been presented. Through the calculation of coupled wave theory, three different lengths of TE_{02}-TE_{01} mode converters of 65.43 mm (4 segments), 119.3 mm (6 segments) and 136 mm (8 segments) are optimized, the mode conversion efficiencies of these mode converters are 91.8–94%, 93–95%, and 95–98.78%, in the design frequency band 215–225 GHz. According to the same optimization method, the TE_{03}-TE_{02} mode converter is designed with a conversion efficiency higher than 95% in the operating frequency band and a conversion efficiency of 98.44% at 220 GHz. Its length is 92 mm (8 segments). Because the length of the mode converter is clearly limited, this paper selects the TE_{02}-TE_{01} mode converter with a length of 65.43 mm (4 segments) and the TE_{03}-TE_{02} mode converter with 92 mm (8 segments) for simulation and experimental verification. 3D simulation software was used to model and simulate the two converters and the TE_{03}-TE_{01} mode converter composed of them. The simulation result curves of the three mode converters are in good agreement with the theoretical calculation results, and there are only varying degrees of frequency deviation. The frequency deviation of the 4-stage TE_{02}-TE_{01} mode converter can be ignored. The frequency deviations of the TE_{03}-TE_{02} mode converter and the TE_{03}-TE_{01} mode converter are 2 GHz and 3 GHz. In this paper, the mode of the input and output ports of the mode converter is measured by means of electric field scanning. When the input mode purity is 92% in TE_{01} mode, the mode purity of TE_{03} mode output of the mode converter is 82%, and the transmission loss of the measured mode converter is low. The measurement results further verify the correctness of the theoretical and simulation results, and the prepared mode converter meets the experimental requirements of our gyrotron. In general, this paper verifies the feasibility of the taper cascaded over-mode circular waveguide mode converter from three aspects of theory, simulation, and experiment. This type of mode converter is easy to prepare, which makes it an effective alternative for high frequency curvilinear waveguide mode converter.

Author Contributions: T.Y. and W.F. contributed to the overall study design, analysis, computer simulation, and writing of the manuscript. X.G., D.L., X.Y. and Y.Y. provided technical support and revised the manuscript. All authors have read and agreed to the published version of the manuscript.

Funding: This work was supported in part by National Key Research and Development Program of China under 2019YFA0210202, in part by the National Natural Science Foundation of China under Grant 61971097 and 6201101342, and in part by the Terahertz Science and Technology Key Laboratory of Sichuan Province Foundation under Grant THZSC201801.

Institutional Review Board Statement: Ethical review and approval were waived for this study, due to this study not involving humans or animals.

Informed Consent Statement: Informed consent was obtained from all subjects involved in the study.

Data Availability Statement: Data sharing not applicable.

Acknowledgments: The authors gratefully acknowledge Yin Huang and Weirong Deng for their kind assistance on engineering design and assembling

Conflicts of Interest: The authors declare no conflict of interest.

References

1. Gilmour, A.S., Jr. *Principles of Klystrons, Traveling Wave Tubes, Magnetrons, Cross-Field Ampliers, and Gyrotrons*; Artech House: Norwood, MA, USA, 2011.
2. Rzesnicki, T.; Piosczyk, B.; Kern, S.; Llly, S.; Jin, J.; Samartsev, A.; Thumm, M. 2.1: Experiments with the European 2 MW coaxial-cavity pre-prototype gyrotron for ITER. In Proceedings of the 2010 IEEE International Vacuum Electronics Conference (IVEC), Monterey, CA, USA, 18–20 May 2010; pp. 27–28. [CrossRef]
3. Lan, F.; Yang, Z.Q.; Shi, Z.J. Study on $TE\_(0n)$ nonuniform ripple-wall mode converter in circular waveguide. *Acta Phys. Sinica.* **2012**, *61*, 155201. [CrossRef]
4. Liu, D.W.; Wang, W.; Zhuang, Q.; Yan, Y. Theoretical and Experimental Investigations on the Quasi-Optical Mode Converter for a Pulsed Terahertz Gyrotron. *IEEE Electron Device Lett.* **2015**, *36*, 195–197. [CrossRef]
5. Thumm, M.; Erckmann, V.; Janzen, G.; Kasparek, W.; Wilhelm, R.; Schüller, P.G.; Müller, G. Generation of the Gaussian-like HE11 mode from gyrotron TEOn mode mixtures at 70 GHz. *Int. J. Infrared Millim. Waves* **1985**, *6*, 459–470. [CrossRef]
6. Thumm, M. Computer-aided analysis and design of corrugated TE11 to HE11 mode converters in highly overmoded waveguides. *Int. J. Infrared Millim. Waves* **1985**, *6*, 577–597. [CrossRef]
7. Doane, J.L. Mode converters for generating the HE 11 (gaussian-like) mode from TE 01 in a circular waveguide. *Int. J. Electron.* **1982**, *53*, 573–585. [CrossRef]
8. Kumric, H.; Thumm, M.; Wilhem, R. Optimization of mode converters. *Int. J. Electron.* **1988**, *64*, 1109–1133.
9. Doane, J.L. Polarization converters for circular waveguide modes. *Int. J. Electron.* **1986**, *61*, 1109–1133. [CrossRef]

10. Guan, X.; Fu, W.; Lu, D.; Yan, Y.; Yang, T.; Yuan, X. Experiment of a High-Power Sub-THz Gyrotron Operating in High-Order Axial Modes. *IEEE Trans. Electron Devices* **2019**, *66*, 2752–2757. [CrossRef]
11. Yu, X.; Deng, J.; Cao, W.; Li, S.; Gao, X.; Yan-Nan, J. Method for Synthesis of TE01-TE11 Mode Converter for Gyrotron by the NURBS Technique. *IEEE Trans. Microw. Theory Tech.* **2015**, *63*, 326–330. [CrossRef]
12. Chittora, A.; Mukherjee, J.; Singh, S.; Sharma, A. Dielectric loaded tm01 to te11 mode converter for S-band applications. *IEEE Trans. Dielectr. Electr. Insul.* **2015**, *22*, 2057–2063. [CrossRef]
13. Dhuda, H.V.; Patel, P.N.; Pandya, H.B. Design and development of the W-band corrugated waveguide mode converter for fusion plasma experiments. *Int. J. RF Microw. Comput. Eng.* **2019**, *30*. [CrossRef]
14. Li, H.; Thumm, M. Mode coupling in corrugated waveguides with varying wall impedance and diameter change. *Int. J. Electron.* **1991**, *71*, 827–844. [CrossRef]
15. CST STUDIO SUITE™, CST AG, Germany. Available online: www.cst.com (accessed on 3 January 2021).
16. Guan, X.; Fu, W.; Yan, Y. Demonstration of a High-Order Mode Input Coupler for a 220-GHz Confocal Gyrotron Traveling Wave Tube. *J. Infrared Millim. Terahertz Waves* **2017**, *39*, 183–194. [CrossRef]

MDPI

St. Alban-Anlage 66

4052 Basel

Switzerland

Tel. +41 61 683 77 34

Fax +41 61 302 89 18

www.mdpi.com

Electronics Editorial Office

E-mail: electronics@mdpi.com

www.mdpi.com/journal/electronics